TEACHING MATHEMATICS:

A Sourcebook
of Aids, Activities,
and Strategies

TEACHING MATHEMATICS:

A Sourcebook of Aids, Activities, and Strategies

MAX A. SOBEL and EVAN M. MALETSKY

Montclair State College

PRENTICE-HALL, INC., *Englewood Cliffs, New Jersey*

Library of Congress Cataloging in Publication Data

SOBEL, MAX A
    Teaching mathematics.

    Includes bibliographies.
    1. Mathematics—Study and teaching (Secondary)
I. Maletsky, Evan M., joint author. II. Title.
QA11.S663   510'.7'12   74-10874
ISBN   0-13-894139-4
ISBN   0-13-894121-1 (pbk.)

© 1975 by Prentice-Hall, Inc., Englewood Cliffs, N. J.

Printed in the United States of America

10 9 8 7 6 5 4 3 2 1

PRENTICE-HALL INTERNATIONAL, INC., *London*
PRENTICE-HALL OF AUSTRALIA, PTY. LTD., *Sydney*
PRENTICE-HALL OF CANADA, LTD., *Toronto*
PRENTICE-HALL OF INDIA PRIVATE LIMITED, *New Delhi*
PRENTICE-HALL OF JAPAN, INC., *Tokyo*

Photographs that appear throughout the book
were taken by
Kerry D. Maletsky

CONTENTS

PREFACE

Teaching Mathematics: Aids, Activities, and Strategies is designed for use by the mathematics teacher and the teacher in training. It treats the art of teaching through a series of motivational ideas suitable for many grade levels and abilities, a discussion of various recreational activities appropriate for the classroom, and a collection of laboratory experiments planned for individual or class use. It contains instructions for many inexpensive aids and models useful in teaching, suggestions for developing audio-visual facilities, and extensive listings of resources and references.

Teaching Mathematics was written at the request of many teachers and students of mathematics who have attended lectures, courses, and workshops given by the authors. Although these teachers and students agreed with the philosophy that it is necessary to properly motivate students, that discovery methods should be employed where possible, that laboratory techniques are effective, and that the use of audio-visual aids should be encouraged, they constantly deplored the lack of sufficient reference materials in a single convenient source. As a result, the authors have collaborated to provide this collection of teaching aids, activities, and strategies suitable in elementary and secondary classes. It is by no means exhaustive on the subject; rather it is suggestive of the types of teaching methods and materials that can be effectively used to improve the teaching of mathematics.

Much of the material that appears in this book has been presented and tested in a variety of ways. Some of the content, especially from the first chapter, is based on talks given by the authors at various state and national professional meetings throughout the country. Many of the ideas concerning the use of multi-sensory aids have been used by the authors in various undergraduate and graduate courses in mathematics education taught at Montclair State College and elsewhere. Most of the laboratory and discovery activities suggested have been used by the authors with many junior and senior high school mathematics classes that they have taught personally.

A leading educator once said that we have done a reasonably good job in the past decade in teaching better mathematics, but that it is now time to learn to teach mathematics better. It is the fond hope of the authors that the materials and ideas presented in this book will enable teachers of mathematics to improve their pedagogical skills and thus better motivate students to learn their subject matter. To their many students —past, present, and future—the authors fondly and encouragingly dedicate this volume as one small contribution to the big task of teaching mathematics better.

MAX A. SOBEL

EVAN M. MALETSKY

TEACHING MATHEMATICS:

A Sourcebook
of Aids, Activities,
and Strategies

CHAPTER ONE

THE ART
OF TEACHING

Although the source is unknown, the authors once heard the job description of a teacher summarized by these three statements:

> A teacher must know his stuff.
> He must know the pupils he intends to stuff.
> Above all, he must stuff them artistically.

It is the last item of this list that we are primarily concerned with in this chapter, and indeed throughout this book. We all wish to improve our artistry in "stuffing" our students with appropriate, contemporary mathematics; this artistry is important when we attempt to motivate our instruction and to challenge the very many reluctant learners that cross our paths daily.

Every teacher has his or her own "bag of tricks" that may work effectively. However, all conscientious teachers are constantly on the search for new ideas and techniques to adopt in their classrooms. Therefore, it is hoped that the collection of items presented in this chapter may serve to provide some additional procedures that mathematics teachers will find useful in their daily teaching.

Without making an effort to enumerate any items in magnitude of importance, this chapter lists a number of guiding principles that the authors believe are the essential ingredients of any mixture that is to produce artistry in teaching. Many of these principles will then be elaborated upon in later chapters.

1.1 Ask interesting and exciting questions

An interesting question can often serve as one of the most effective ways to start or end a class. A question is posed and students are given an opportunity to guess and debate the answer. Then, with teacher guidance, appropriate methods are considered for the solution of the problem. Of course the question is so designed that its solution requires the class to employ mathematical methods appropriate to the curriculum and level of instruction at hand.

Consider, for example, a seventh-grade class that happens to be studying a unit on our decimal system of notation. The teacher wishes to provide some review of fundamental computation, and also hopes to develop an appreciation of the meaning of very large numbers. An interesting question that could be used to start the class is the following:

> "I've just decided that I would like to count to 1 million. I plan to do so at the rate of one number per second until I finish, that is, 1, 2, 3, 4, 5, and so on. If I don't stop to do anything else until I'm finished, how long will it take me?"

Of course there will always be one bright student to give the answer as 1 million seconds! However, the teacher then asks for the answer in more commonly understood units of time, such as days, weeks, months, or years. At this point it is extremely important to allow students to guess before they compute. A heated discussion among students concerning their guesses is the best way to motivate them to perform the computation necessary to provide the correct answer.

A word of caution should be inserted here. Occasionally one meets a class where most attempts at motivation seem to fail. After discussion of the various guesses for the time it takes to count to 1 million, the teacher must lead the class to discover the computational method used to find the correct answer. If the class is not really interested at this point in the correct answer, there is hardly any purpose in pursuing the matter further. It is for this reason that it is important to generate sufficient discussion in advance so that students are eager to learn the solution.

Many interesting questions can be used to stimulate discussion and motivate computation. Here are a few; in each case the important approach is to guess and discuss first and then compute.

1. Consider 1 million pennies piled one penny on top of another. How high would the pile reach? As high as the ceiling? As high as the

flagpole? As high as the Empire State Building? As high as the moon?
2. One million $1 bills are placed end to end on the ground. How far
would they reach? Across a football field? Across the state? Across the
United States? Around the world?
3. I'm going to snap my fingers. One minute later I'll snap them again.
Then 2 minutes later I'll snap them once more. Then I'll wait 4 minutes
to snap them; then 8 minutes, 16 minutes, and so on. Each time I
double the number of minutes in the interval between snaps. At this
rate, how many times will I snap my fingers in one year?

As a final example, consider an eighth-grade class that is studying a
unit on measurement. An interesting question that could be used to start
the class might be the following: "Look at our classroom. Do you think
we could fit 1 million basketballs into this room? How about 1 million
baseballs? One million table-tennis balls? One million marbles? One million pennies?"

For a classroom of average size the guess is often between 1 million
table-tennis balls and 1 million pennies. After time for discussion, the
class is asked to determine a procedure by which to approximate the correct answer. One possibility might be to bring an empty shoebox to class
and fill it up with table-tennis balls. Then use a tape measure and approximate the volume of the room. By comparing the volume of the room with
the volume of the shoebox, one can obtain a fair approximation to the
number of table-tennis balls that could fit into the room.

Another question that could be used to serve the same purpose, or
could be given as an assignment, is to determine the approximate weight
of 1 million pennies. Questions of this type generally serve to set the stage
for interesting class discussions that make students look forward to attend-
ing their mathematics class . . . a place where exciting things happen!

1.2 Make provisions for student discovery

Discovery techniques can be effectively used to stimulate and maintain interest in mathematics. Furthermore, such approaches help to develop the
type of creativity and originality that is important for a student's future
success in mathematics.

The famous French mathematician René Descartes (1596–1650) con-
cluded his book *La Geometrie* with this comment:

> I hope that posterity will judge me kindly, not only as to the things
> which I have explained, but also as to those which I have inten-
> tionally omitted so as to leave to others the pleasure of discovery.

In our desire to impart knowledge to our students, it is important that we do not fail to make provisions for them to participate in and to enjoy this all-important pleasure.

An entire book could well be written on just this one topic. Indeed, many mathematics textbooks make the claim that they feature a discovery approach. Actually, there are two different types of discovery approaches. At its purist level we have *creative discovery,* wherein a teacher presents a situation to a class and allows the students to explore on their own, using their intuition and past learning, with little or no guided direction. Such an approach is especially well suited for the gifted student and provides him with the type of experience that is necessary for later independent research.

As an example of such an approach, consider an eleventh- or twelfth-grade class that has already studied a unit of work on arithmetic and geometric series. The teacher then proceeds to tell the class about the infinite tree. This tree looks like any other tree but grows in a very interesting manner. The first day the tree grows 1 foot. The second day two new branches grow, each $\frac{1}{2}$ foot in length and at right angles to each other. The next day two new branches appear at each terminal point, again at right angles to one another but only $\frac{1}{4}$ foot in length. This continues forever! This is what the tree looks like during the first four days of growth:

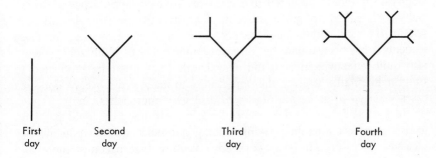

First   Second   Third   Fourth
day   day   day   day

At this point, in the typical mathematics class, the teacher would ask the students to prove or find certain relationships. Using a creative discovery approach, the teacher terminates the story of the infinite tree by asking the students to discover whatever they can about it, offering no further clues or direction. Among the many interesting items that students may discover, using only right-triangle relationships and knowledge of geometric series, is that the tree has a limiting height of $\dfrac{4+\sqrt{2}}{3}$ feet and a limiting breadth of $\dfrac{2(\sqrt{2}+1)}{3}$ feet.

A similar problem is that of the infinite snowflake. It hits the ground in the shape of an equilateral triangle. Thereafter, each second a new equilateral triangle emerges in the middle third of each side, continuing forever. This is what the first three stages look like:

Once again the student is asked to discover whatever he can about the infinite snowflake, with no teacher direction. Among the many interesting facts that can be discovered is that there is no limiting perimeter to this curve but that there is a limiting area, $\frac{2}{5} n^2\sqrt{3}$ square units, where n is the length of the side of the original equilateral triangle. More on this interesting curve can be read in *Mathematics and the Imagination,* by Edward Kasner and James R. Newman (New York: Simon and Schuster, 1940; paperback ed., 1963).

Again it should be noted that creative discovery approaches are especially well suited to gifted students. The average or slow learner can seldom function without any teacher direction at all. However, for these students it has been found that discovery approaches of a different type are also quite effective; indeed, the slow learner can often be motivated to study mathematics if given the opportunity to make a discovery. For such cases we need to make use of *guided discovery* approaches.

In guided discovery the teacher leads a class along the right path, rejecting incorrect attempts, asking leading questions, and introducing key ideas as necessary. It is a cooperative venture that becomes more and more exciting as one approaches a final result.

The following items should serve to illustrate this approach.

1. A class is asked to find this sum:

$$\frac{1}{1 \cdot 2} + \frac{1}{2 \cdot 3} + \frac{1}{3 \cdot 4} + \cdots + \frac{1}{99 \cdot 100}$$

The task appears to be impossible. The teacher suggests that one approach to problem solving is to consider a small part of the problem at a time. Thus the class is led to consider the first term, the first two terms, the first three terms, and so on:

$$\frac{1}{1\cdot2} = \frac{1}{2}$$

$$\frac{1}{1\cdot2} + \frac{1}{2\cdot3} = \frac{2}{3}$$

$$\frac{1}{1\cdot2} + \frac{1}{2\cdot3} + \frac{1}{3\cdot4} = \frac{3}{4}$$

At this point the teacher asks the class to guess the sum of the first four terms, pointing out the pattern as necessary. Hopefully there will be members of the class to guess that it will be $\frac{4}{5}$. This answer is confirmed by actual computation. Finally, the class should be ready to guess that the answer to the given problem is $\frac{99}{100}$. Of course, it is important to point out that this is just a conjecture and not a proof.

The series can be proved to have a sum of $\frac{99}{100}$ by elementary methods. First we need to recognize these relationships, which also lend themselves to discovery approaches:

$$\frac{1}{1\cdot2} = \frac{1}{2} = \frac{1}{1} - \frac{1}{2} \qquad \frac{1}{2\cdot3} = \frac{1}{6} = \frac{1}{2} - \frac{1}{3} \qquad \frac{1}{3\cdot4} = \frac{1}{12} = \frac{1}{3} - \frac{1}{4}$$

Then write the given series as follows:

$$(1 - \frac{1}{2}) + (\frac{1}{2} - \frac{1}{3}) + (\frac{1}{3} - \frac{1}{4}) + \cdots + (\frac{1}{98} - \frac{1}{99}) + (\frac{1}{99} - \frac{1}{100})$$

Finally, note that every term except the first and last subtract out, giving the sum $1 - \frac{1}{100}$, or $\frac{99}{100}$.

2. A man earns $1 on the first day of a job. The second day he is paid $2, the third day $4, the fourth day $8, and so on. Each day his salary is twice that of the preceding day. He plans to stay on the job for 15 days and wishes to know what his total earnings will be. The specific amounts for each day could be listed and their sum found, but the class is asked to discover a method for finding the sum without actual addition of each of the 15 amounts.

Again one recalls that it is frequently helpful to consider a small portion of a problem at a time. That is, let us consider what happens if he should only work for 3 days, then for 4 days, and then for 5 days.

For 3 days	*For 4 days*	*For 5 days*
1	1	1
2	2	2
4	4	4
Total 7	8	8
	Total 15	16
		Total 31

At this point the class is asked to observe relationships. With teacher assistance they should note that the total salary for 3 days is $1 less than the salary for the fourth day; the total for 4 days ($15) is $1 less than the salary for the fifth day ($16); the total for 5 days ($31) is $1 less than the salary for the sixth day ($32). The discovery is thus made that the total for n days is one less than the $(n + 1)$ term. The total salary for 15 days is given as follows:

$$(1 + 2 + 4 + 8 + 16 + 32 + 64 + 128 + 256 + \\ 512 + 1024 + 2048 + 4096 + 8192 + 16{,}384)$$

The sum is 1 less than the next term; $2 \times 16{,}384 - 1 = 32{,}767$. His total salary is $32,767.

3. As has already been stated, it is important that students recognize that a conjecture is not a proof and that without a proof there is no guarantee that a pattern will continue forever. Thus it is worthwhile to occasionally display a pattern that fails after a certain point. One of the most dramatic ones concerns the number of regions into which a circle can be divided by connecting points on the circle. Consider the following apparent pattern:

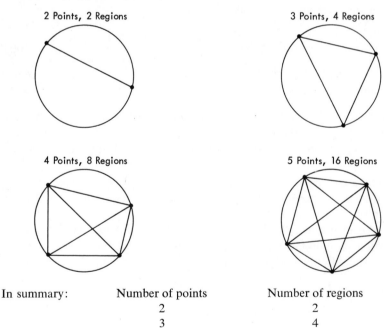

2 Points, 2 Regions

3 Points, 4 Regions

4 Points, 8 Regions

5 Points, 16 Regions

In summary:

Number of points	Number of regions
2	2
3	4
4	8
5	16

How many regions should then be expected for six points connected in all possible ways?

The apparent answer is 32. However, much to everyone's surprise, the maximum number of regions possible with six points proves to be 31!

4. Algebra classes often prove to be the place where one seldom provides opportunities for student discovery because of the undue emphasis on mechanics. Nevertheless, much of the beauty of mathematics centers around the power of generalization, and this can best be developed in algebra courses. Consider, for example, the Fibonacci sequence:

$$1, 1, 2, 3, 5, 8, 13, 21, 34, 55, 89, 144, 233, \ldots$$

In this sequence, each term after the second is the sum of the two preceding terms. Here is a flow chart that can be used quite effectively to generate the first 10 terms of this sequence:

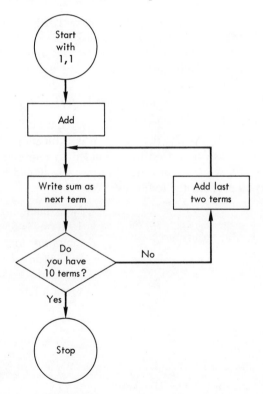

A Fibonacci sequence can be generated by starting with any two numbers and thereafter finding each term as the sum of the two preceding terms. This sequence has many interesting properties, one of which is that the sum of the first 10 terms of a Fibonacci sequence is equal to 11 times the seventh term. This affords the opportunity for an interesting classroom demonstration and a worthwhile lesson in discovering a generalization.

A student is asked to write any two numbers on the board. Assume that he writes 4 and 9. The class then helps to list the first 10 terms of a Fibonacci sequence that starts with these two numbers. When the class reaches the seventh term, 92, the teacher mentally multiplies this by 11 and writes 1012 on the side of the board. The class continues through the tenth term, and the students are then asked to find the sum. They are amazed to find that the sum is a number that the teacher had written down before they ever completed writing the sequence!

1:	4	
2:	9	
3:	13	
4:	22	
5:	35	
6:	57	
7:	92	$\ldots 11 \times 92 = 1012$
8:	149	
9:	241	
10:	390	

$1012 =$ sum of first 10 terms

Finally the class is asked to try and discover the teacher's method. A hint may be given, if necessary, that the sum is a function of the seventh term. Possibly the class may need to be guided to represent the first 10 terms of a Fibonacci sequence in general terms; this should be all the hint that is necessary to produce the following:

1:	a
2:	b
3:	$a + b$
4:	$a + 2b$
5:	$2a + 3b$
6:	$3a + 5b$
7:	$5a + 8b$
8:	$8a + 13b$
9:	$13a + 21b$
10:	$21a + 34b$

Note that the sum of the first 10 terms is $55a + 88b$. By factoring, this is equivalent to $11(5a + 8b)$, which proves to be 11 times the seventh term.

$55a + 88b =$ sum of first 10 terms

1.3   Make mathematics come alive

Too many of our students think of mathematics as a very dull subject, and they picture mathematicians as hermits who spend their lives buried in mountains of figures. One interesting way to make mathematics come alive is to make frequent use of historical items that help to show that mathematicians are human beings, with mortal weaknesses and interests. Books on the history of mathematics are excellent sources for such items, many of which can be appropriately used to introduce or to supplement particular topics in the classroom.

As an example, consider the story told of the German mathematician Karl Friedrich Gauss (1777–1855), considered by many to have been the greatest mathematician of all times. It is claimed that young Gauss was a precocious youngster with a born flair for mathematics. To keep him suitably occupied, his teacher in elementary school told Gauss to write the numbers from 1 to 100 and to find their sum. In a flash Gauss is supposed to have given the answer as 5050. It is further claimed that he did so by mentally recognizing this pattern:

$$1 + 2 + 3 + \cdots\cdots\cdots\cdots\cdots + 98 + 99 + 100$$
$$3 + 98 = 101$$
$$2 + 99 = 101$$
$$1 + 100 = 101$$

Since there are 50 pairs with sum of 101, the total sum is $50 \times 101 = 5050$! Today we use the formula $S = \dfrac{n(n + 1)}{2}$ for the sum of the first n successive counting numbers.

The story may or may not be true, but it serves as an interesting anecdote to use in the junior high school grades, or in a senior algebra class when introducing a unit on arithmetic series. Gauss is claimed to have said "Mathematics is the queen of the sciences, and arithmetic is the queen of mathematics." This and other quotes by famous mathematicians serve as effective bulletin board material.

Unsolved and impossible problems are usually of great interest to mathematics students. Every teacher of geometry has experienced his share of would-be angle trisectors, students who attempt the impossible. An interesting anecdote can be told about an unsolved problem known as Fermat's Last Theorem. Pierre de Fermat (1601?–1665) claimed that there do not exist positive integers, x, y, and z such that $x^n + y^n = z^n$, for $n > 2$. (Note that for $n = 2$, we have many examples of the equality, such as $3^2 + 4^2 = 5^2$ and $5^2 + 12^2 = 13^2$.)

Fermat claimed that he had discovered a proof of the impossibility of this relation but that the margin of his book was too narrow to include it. To this date no mathematician has been able to prove the impossibility of this problem. It is therefore felt that Fermat believed he had a proof but that there was undoubtedly some flaw in it; but of course that is just a hypothesis.

In the early 1900s, a professor of mathematics in Darmstadt, Germany,

Paul Wolfskehl, spent many long hours in a vain attempt to prove Fermat's Theorem. He was also disappointed in love, and so decided to end his life. Being a methodical man, he wrote a suicide note specifying the date and hour that he would commit the act. Within a few hours of the appointed time he decided to occupy his final hours by one last look at Fermat's Theorem. It is said that he became so engrossed once again in the problem that his appointed suicide hour passed by unnoticed, whereupon he tore up his note and began life with renewed energy. When Wolfskehl finally died in 1908 he left a will that provided 100,000 marks to the first one to prove Fermat's Theorem, a sum long since depleted by inflation.

There are many ways that the creative teacher can use such a story to generate interest in a classroom. It can be used in a ninth- or tenth-grade class about to begin a study of the Pythagorean theorem. The teacher prepares a newspaper-like sign on the blackboard or on the bulletin board that states "MATHEMATICS SAVES A LIFE!" The story of Wolfskehl serves as an exciting introduction to the unit of study at hand.

Surprisingly, most secondary students of mathematics appear to be quite interested in historical items. Ancient systems of computation, for example, can be used quite effectively to enliven a class and serve well as an introduction to modern approaches. We shall consider two specific examples, one arithmetic and the other algebraic.

The ancient Egyptian method of duplation and mediation for multiplication fascinates most students. To multiply two numbers, such as 25×32, one doubles one of the factors while halving the other. When dividing by 2, all remainders are discarded, as shown below:

Divide by 2	*Multiply by 2*	
(25)	**32**	Note that $25 \div 2$ is written
12	64	as 12, and $3 \div 2$ is written
6	128	as 1 since remainders are
(3)	**256**	discarded.
(1)	**512**	

Answer: $25 \times 32 = 32 + 256 + 512 = 800$

Next, all the odd numbers in the first column are circled. The numbers opposite these in the second column (32, 256, 512) are added. Their sum, 800, represents the product 25×32. Junior high school students enjoy trying this method for various products, and it serves well as a means of motivating computation. A discussion of binary notation can be used to justify the procedure.

Classes in algebra will enjoy the ancient methods of solving equations. The "rule of false position" was known to the early Egyptians and was transmitted in the Middle Ages to Europe from India. To solve an equation, one first takes a guess. Consider the equation

$$x + \frac{x}{7} = 24$$

and let us guess that the answer is 7. Using 7 in place of x we have

$$7 + \frac{7}{7} = 8$$

But the given equation has been set equal to 24, which is 3 times 8. Therefore, the correct solution must be 3 times 7, or 21.

The "rule of double false position" is more complicated, but it can be used very effectively to introduce or motivate a unit on the solution of pairs of linear equations. Here one begins by taking two guesses, x_1 and x_2. These guesses are then substituted in the given equation and the results are noted as r_1 and r_2. The correct solution is then found by substitution in this formula:

$$x = \frac{r_1 x_2 - r_2 x_1}{r_1 - r_2}$$

Consider the equation $3x - 12 = 0$ with guesses $x_1 = 2$ and $x_2 = 5$:

$$\text{For } x_1 = 2: r_1 = 3(2) - 12 = -6$$

$$\text{For } x_2 = 5: r_2 = 3(5) - 12 = 3$$

By substitution in the formula we have

$$x = \frac{(-6)(5) - (3)(2)}{-6 - 3} = \frac{-36}{-9} = 4$$

After applying the formula to several problems to be assured that it works, the class can be asked to justify it and can be led to do so by considering the general linear equation $ax + b = 0$:

$$\text{For } x = x_1: \quad \begin{cases} ax_1 + b = r_1 \\ ax_2 + b = r_2 \end{cases}$$
$$\text{For } x = x_2:$$

Now we also know that the solution of the general equation $ax + b = 0$ is $x = -b/a$. Thus all that is needed to verify the given formula for the rule of false position is to use the linear equations given and find $-b/a$ in terms of x_1, x_2, r_1, and r_2.

There are many good sources of information for historical data and anecdotes. Several recommended references follow:

BELL, ERIC T. *Men of Mathematics*. New York: Dover Publications, Inc., 1937.

DAVIS, PHILIP J. and CHINN, WILLIAM G. *3.1416 and All That*. New York: Simon and Schuster, 1969.

EVES, HOWARD. *In Mathematical Circles*. Boston: Prindle, Weber & Schmidt, Inc., 1969.

————. *An Introduction to the History of Mathematics*. New York: Holt, Rinehart and Winston, Inc., 1969.

KLINE, MORRIS. *Mathematical Thought from Ancient to Modern Times*. New York: Oxford University Press, Inc., 1972.

National Council of Teachers of Mathematics. *Historical Topics for the Mathematics Classroom* (Thirty-first Yearbook). Washington, D.C.: N.C.T.M., 1969.

1.4   Start or end the period with something spectacular

This bit of advice is not easy to follow inasmuch as the supply of truly spectacular items is limited. However, through the years a teacher should be able to collect a supply of "gimmicks" to help enliven a class. Having something special at the start of a period occasionally has the result of making students truly want to come to the mathematics class. Using such items at the end results in having students regret having the period come to a close, and walking out talking about the exciting things that happen in their mathematics class! In any event the appropriate motivation to study mathematics is the desirable outcome that should result.

Many spectacular things can be done by embellishing upon rather simple mathematical tricks or patterns. Several such examples are provided here, others are inserted throughout this text, and still others may be found by careful search of the available literature.

1. A student is asked to go to the board and write a two-digit number between 50 and 100, without the teacher's seeing it. He is then told to add 76 to this number. Next he is told to cross out the digit in the hundred's place of his answer and add this to the remaining two-digit number. Finally, he is asked to subtract his result from his original number. These are the steps if the student were to begin with the number 83:

Original number:	83
Add 76:	$+\ 76$
	$\overline{159}$
Cross out and add:	$\cancel{1}59;\quad 59 + 1 = 60$
	83
Subtract from original number:	$-\ 60$
Result is 23:	$\overline{23}$

The interesting thing about this trick is that the final outcome will always be 23, regardless of the number selected by the student, provided that the stated steps are followed. However, it is certainly not very spectacular to conclude by announcing that the final outcome is 23, as interesting as this may be. A far more dramatic approach is the following.

Before coming to class the teacher uses the edge of a damp piece of soap and writes 23 on the back of his hand. When this dries it becomes completely invisible to the student. In class, after completing the puzzle, the teacher proceeds to ask a member of the class to write the final outcome on a piece of paper and fold it. He then carefully burns the paper in some suitable receptacle and waits until the ashes cool off. Finally he picks up the ashes and wipes this on the back of his hand, whereupon, as if by magic, the number 23 is clearly outlined for all to see! It has been the experience of the authors that this trick is one that students continue to talk about at great length for many weeks.

2. Not quite as spectacular, but nevertheless an interesting experiment, is to walk into class and place on the board the "magic number" for the day:

$$12,345,679$$

When the class asks what is magic about this number, the teacher proceeds to ask row 1 to multiply the number by 9. Row 2 is to multiply the same number by 18 (2×9), row 3 by 27 (3×9), row 4 by 36 (4×9), and row 5 by 45 (5×9). The following interesting products are obtained:

$$9 \times 12,345,679 = 111,111,111$$
$$18 \times 12,345,679 = 222,222,222$$
$$27 \times 12,345,679 = 333,333,333$$
$$36 \times 12,345,679 = 444,444,444$$
$$45 \times 12,345,679 = 555,555,555$$

This same pattern continues for the remaining multiples of 9 through 81.

Obviously there are many ways in which any of these examples may be used. For example, as an alternative approach the teacher asks the class if anyone has a favorite or lucky number between 1 and 10. If a student responds that 3 is his lucky number he is asked to multiply 12,345,679 by 27 and report on the result; the student who offers 7 as his lucky number is told to multiply 12,345,679 by 63; and so forth. The surprise on the students' faces as they complete the multiplication is rewarding and can be contagious for the rest of the class as well!

3. Fallacies are generally of interest to most secondary mathematics students. An interesting start to a class period can be effected by the

teacher announcing that he has just discovered a proof that $1 = 2$. This is a standard fallacy that appears in many texts:

Let $a = b$
Then $a \cdot a = a \cdot b;$ that is, $a^2 = ab$
By subtraction, $a^2 - b^2 = ab - b^2$
By factoring, $(a - b)(a + b) = b(a - b)$
By division, $a + b = b$
Thus, $b + b = b$ since $a = b$
Finally, $2b = b$ and $2 = 1$

Of course, the fallacy lies in the fact that we divided by 0 in the form of $a - b$. This can then lead to a well-motivated discussion of why division by zero is not permissible.

1.5   Make effective use of multisensory aids

Actually this item is what this book is all about. The chapters that follow will provide the reader with numerous suggestions for aids that can be used in the mathematics classroom to promote learning. The major emphasis throughout will be on aids that the teacher can readily provide, with a minimum of time and effort, as opposed to costly commercial aids. Although many of the latter are especially good, most teachers of mathematics are more receptive to the use of aids if they can be quickly assembled and used.

What uses can be found for a strip of paper off a roll of register tape? Some teachers would use it to bring a googol to class. A googol can be written as 1 followed by 100 zeros, or simply as 10^{100}. The number doesn't sound too big when you talk about it, but the student can begin to appreciate just how large it is when the teacher pulls a strip of paper from his pocket and unrolls the 100 zeros clear across the classroom. By comparison, 1 million looks trivial.

GOOGOL   10,000,000,000,000,000,000,000,000,000,000,000,000,
          000,000,000,000,000,000,000,000,000,000,000,000,
          000,000,000,000,000,000,000

Patterns for representing infinite repeating and non-repeating decimals can also be effectively displayed this way.

repeating   .3113113113113113113113113113113113113113311 ...

non-repeating .31133113331133331133333113333331133333311 ...

When discussing the nonrepeating nature of the decimal representation for the irrational number π, many students nod in agreement but do not really believe that it *never* repeats. The message is much more impressive and lasting when the students can actually see and study 100 or so of the digits for repeating patterns. All the teacher needs is a strip of paper and the time to copy the digits down.

$$\pi = 3.14159\ \ 26535\ \ 89793\ \ 23846\ \ 26433\ \ 83279\ \ 50288\ \ 41971$$
$$69399\ \ 37510\ \ 58209\ \ 74944\ \ 59230\ \ 78164\ \ 06286\ \ 20899$$
$$86280\ \ 34825\ \ 34211\ \ 70679\ \ .\ .\ .$$

A listing of the first 4000 decimals in the expansion of π can be found in *The Lore of Large Numbers,* by Philip J. Davis (New York: Random House, Inc., 1961, pp. 72–73).

A sheet of paper can also be a simple, handy aid for the mathematics teacher. Paper-folding activities frequently stimulate interest in geometry as well as provide challenging problems. For example, at the end of the study of equilateral triangles, students can explore methods for constructing one by folding paper using the shorter edge as one side:

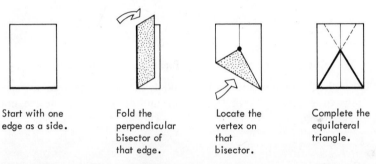

| Start with one edge as a side. | Fold the perpendicular bisector of that edge. | Locate the vertex on that bisector. | Complete the equilateral triangle. |

Another activity with paper might involve an informal proof, as in this illustration of the Pythagorean theorem:

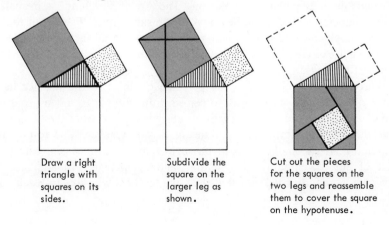

| Draw a right triangle with squares on its sides. | Subdivide the square on the larger leg as shown. | Cut out the pieces for the squares on the two legs and reassemble them to cover the square on the hypotenuse. |

What uses can the teacher find for a piece of string in the mathematics class? He can come to class with a piece stuffed in his pocket and start pulling it out. Each student guesses at the length before they see it all. Obviously, the answers will vary widely. As the rest is pulled out for them to see, students again guess at the length. The answers will still vary and probably more than the students themselves would expect. Next they are asked to guess how many times it will fit around a dollar bill and a basketball. Once the actual length is given, the ends are tied together and students are asked for the dimensions of the largest square and equilateral triangle that can be formed from it. As a last activity, the teacher goes from one student to the next throughout the entire class asking for a different set of dimensions for a rectangle or an isosceles triangle that can be formed with it. With a simple piece of string, students can become actively involved and motivated to a further study of measurement.

Even a simple flashlight can be an aid when teaching the conic sections. Held at different angles, the circular reflector can project a circle, ellipse, parabola, and hyperbola:

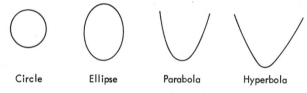

| Circle | Ellipse | Parabola | Hyperbola |

Throughout the remainder of this book the reader will be exposed to numerous other examples of the use of multisensory aids in the mathematics classroom.

1.6 Conclusion

As originally anticipated, a chapter on "The Art of Teaching" is in reality both an impossible and an endless task at the same time! The list of topics that one might include is long and yet can never be all-inclusive. Indeed, an excellent exercise for the reader is to attempt to constantly add to this initial list of items.

Throughout this discussion the personal qualities of the teacher have been bypassed. The easiest way to consider this item is to ask a group of secondary school students to submit an unsigned list of what they consider to be the five most important qualities of a good teacher. The results are predictable. High on the list will be such items as enthusiasm, sincerity, sense of humor, empathy, imagination, and competence. In addition, in their own words, they will say that they like teachers who like to teach.

Thus by our every action we must show our students that we like to teach mathematics. Enthusiasm is contagious, and when the teacher demonstrates a sincere interest in both his students and his subject, his class will seldom ask the embarrassing question, "What good is all this?" The authors of this book sincerely hope that the material in the chapters to follow will be of some help in assisting the teacher of mathematics to teach in an exciting, meaningful, and memorable manner.

Exercises

1. How long will it take to count to 1 million at the rate of one number per second?

2. One million pennies are piled one on top of the other. How high will the pile be?

3. One million $1 bills are placed end to end on the ground. Approximately how far will the bills stretch?

4. A person snaps his fingers. One minute later he snaps them again. He then waits 2 minutes before the next snap and 4 minutes for the following snap. Then he waits 8 minutes. He continues in this manner, each time doubling the interval between snaps. At this rate, how many times will he snap his fingers in one year?

5. In Exercise 4, assume that the person continues snapping his fingers in the manner indicated for a period of two years. How many times will he have snapped his fingers during that time?

6. How many pennies will fit into a man's shoebox of standard size? Guess first, then experiment.

7. For the infinite tree described on page 5, show that the limiting height is $\dfrac{4 + \sqrt{2}}{3}$ feet, and the limiting breadth is $\dfrac{2(\sqrt{2} + 1)}{3}$ feet.

8. For the infinite snowflake described on page 6, show that there is no limiting perimeter but that the limiting area is $\dfrac{2}{5} n^2 \sqrt{3}$ square units, where n is the length of a side of the original triangle.

9. Use Gauss's method, described on page 11, to find the sum of the first n counting numbers.

10. Explain why the "23" trick described on page 14 works.

Activities

1. Prepare a set of five questions that can be used in a junior high school mathematics class to promote guessing, and subsequent discovery of an answer. If possible, test these items in an actual class situation.

2. Repeat Activity 1 for a senior high school mathematics class.

3. Repeat Activity 1 for a ninth-grade general mathematics class of slow learners.

4. Prepare a collection of historical anecdotes about mathematicians that could be used in a mathematics classroom to motivate instruction.

5. Try to make additional discoveries about the infinite tree and snowflake.

6. Try to make additional discoveries about the Fibonacci sequence.

7. Develop a one-period lesson plan for a junior high school mathematics class that features a discovery approach.

8. Repeat Activity 7 for a senior high school mathematics class.

9. Compile a list of different ways that a piece of paper can be used in the mathematics classroom as a visual aid. Do not neglect the use of graph paper.

10. Ask a class of secondary mathematics students to prepare a list of what they consider to be the most important qualities of a good teacher. Then summarize the results in order of the most frequently mentioned items.

References and selected readings

BELL, ERIC T. *Men of Mathematics*. New York: Simon and Schuster, 1965.

BUTLER, CHARLES H., F. L. WREN, and J. H. BANKS. *The Teaching of Secondary Mathematics*. New York: McGraw-Hill Book Company, 1970.

COURANT, RICHARD, and HERBERT ROBBINS. *What Is Mathematics? An Elementary Approach to Ideas and Methods*. New York: Oxford University Press, Inc., 1941.

DAVIS, ROBERT B. *Explorations in Mathematics: A Test for Teachers*. Reading, Mass.: Addison-Wesley Publishing Company, Inc., 1966.

FREMONT, HERBERT. *How To Teach Mathematics in Secondary Schools*. Philadelphia: W. B. Saunders Company, 1969.

HOGBEN, LANCELOT T. *Wonderful World of Mathematics.* Garden City, N. Y.: Doubleday & Company, Inc., 1955.

JOHNSON, DONOVAN A., and G. R. RISING. *Guidelines for Teaching Mathematics.* Belmont, Calif.: Wadsworth Publishing Company, Inc., 1972.

KASNER, EDWARD, and JAMES R. NEWMAN. *Mathematics and the Imagination.* New York: Simon and Schuster, 1940; paperback ed., 1963.

Mathematics Teacher. "Snow Flake Curves" (Apr. 1964), p. 219.

MORITZ, ROBERT E. *On Mathematics: A Collection of Witty, Profound, Amusing Passages About Mathematics and Mathematicians.* New York: Dover Publications, Inc., 1942.

National Council of Teachers of Mathematics. *Historical Topics for the Mathematics Classroom* (Thirty-first Yearbook). Washington, D.C.: N.C.T.M., 1970.

————. *Instructional Aids in Mathematics* (Thirty-fourth Yearbook). Washington, D.C.: N.C.T.M., 1973.

————. *The Teaching of Secondary School Mathematics* (Thirty-third Yearbook). Washington, D.C.: N.C.T.M., 1970, Chap. 6, "The Art of Generating Interest," by Herman Rosenberg.

POLYA, GEORGE. *Mathematical Discovery.* New York: John Wiley & Sons, Inc., 1962.

CHAPTER TWO

MOTIVATIONAL IDEAS

Almost every mathematics educator will agree on the importance of proper motivation in the teaching of mathematics. With the exception of the very few students who seem to have a natural love for the subject, most others need to have their interest stimulated through suitable teaching techniques and procedures. Charles H. Butler, F. L. Wren, and J. H. Banks summarize this point of view quite well in their text *The Teaching of Secondary Mathematics* (New York: McGraw-Hill Book Company, 1970):

> It may be taken as axiomatic that students will work most diligently and most effectively at tasks in which they are genuinely interested. To create and maintain interest becomes, therefore, one of the most important tasks of the teacher of secondary school mathematics. It is also one of the most difficult problems the teacher encounters.

One of the major difficulties of motivation is the frequent inability to locate suitable materials and ideas. Many teachers become so involved with the routines of presenting their subject matter that they lack the necessary time and energy to search for motivational items. Nevertheless, there is an abundant supply of such topics, and this chapter represents a small sampling that will hopefully encourage the reader to search for similar ideas.

The suggestions that follow may be used in the classroom in many different ways. Some are helpful to introduce a specific topic, whereas others are designed to show applications of a unit already studied. Most of them, however, are designed so that the teacher can use them during the first or last few minutes of a period—to start the class off with a bang and/or to keep it listening during those last 5 minutes before the bell. They are all designed to help develop the idea that mathematics can be both interesting and fun. Hopefully, their use will make students look forward to coming to the mathematics class and make them sorry to see the period end!

2.1 Intuitive guessing

George Polya of Stanford University has said that "mathematics in the making consists of guesses." In order to make a discovery, it is first necessary to make a guess . . . and the guess may be hasty; indeed it should be. These guesses then need to be followed by verification, and this is the hard part of mathematics, the proof in support of a guess. But proof is also the least imaginative part of the process; it is the original guess that is the creative part of mathematics.

Most adults are afraid to take a guess for fear of being wrong. On the other hand, most adolescents are quite ready and eager to guess and the teacher should capitalize on this by providing suitable opportunities for intuitive guessing to take place in the classroom.

It is important that students be given sufficient time to formulate guesses and discuss these in class before attempting to find a correct answer through computation. Unless time is taken for this intuitive discussion, the topic serves only to provide a vehicle for computation, and the motivational aspects are lost. Consider, for example, the following question that could be posed at the start or end of a period in a junior high school class: "This piece of paper that I'm holding is .003 inch thick. I'm going to fold it once to give the thickness of two pieces. Then I'll fold it again to have the thickness of four pieces, then again for 8, 16, and so on. Now let's assume that I'm able to continue in this way for 50 folds; how thick would the stack of paper then be?"

As indicated in Chapter 1, if the teacher immediately turns to the task of finding an answer by the necessary computation, the problem loses all interest to the student. Rather, one should urge students to guess and allow the class to establish the range between the lowest and the greatest answers given. After the heated discussion that can be expected to ensue, the teacher can encourage suggestions for the arithmetical procedures needed to find the correct answer. This can be done in class if the question is used at the start of a period to motivate a review lesson on computation, or it can be assigned as homework if the problem is given at the end of the period to stimulate interest in mathematics. The correct answer, over 50,000,000 miles, invariably comes as a big surprise to most students.

An interesting question that can be raised relative to this problem is a guess as to the number of times that one can physically fold a piece of paper. Most students are extremely surprised to find that it is almost impossible to complete more than seven folding operations. Of course, they assume that one might be able to fare better with a larger piece of paper to start with. An interesting experiment can then be performed by attempting the folding with a sheet of newspaper; the results prove to be the same, with seven folds remaining as the maximum number.

The following items are representative of the many questions that are suitable to invite guesses, promote discussion, and motivate the necessary computation to establish answers. Most of the suggestions offered here are particularly worthwhile for use with groups of slow learners in general mathematics inasmuch as these youngsters very much enjoy making guesses and can be stimulated to find answers by appropriate computational procedures. In each case, guess first and then compute.

1. What are the dimensions of a rectangular room that would be large enough to hold 1 million pennies?
2. One million people are lined up, with just an arm's length between each person. How far will this line reach? Will it reach across Pennsylvania? Across the United States? Around the world?
3. A man starts a chain letter. He sends the letter to two people and asks each of them to send copies of it to two other people. These recipients in turn are asked to send copies to two additional people each. Assuming no duplications, how many people will have received copies of this letter after the twentieth mailing?
4. A person snaps his fingers, and then snaps them again 1 second later. He then waits for 2 seconds and snaps them again. Then he waits 4 seconds before the next snap, and 8 seconds before the next one. He continues in this way, each time doubling the interval of seconds between snaps. If he were to continue in this manner for one year, how many times would he snap his fingers during that period of time?
5. A person is offered a job that pays 1 cent the first day, 2 cents the second day, 4 cents the third day, 8 cents the fourth day, and so on. The salary for each day is double the salary for the preceding day. If he were to accept this job for 30 days, what would his total income be for the month?

Interesting guessing activities can be provided in the format of multiple-choice questions. As before, allow students to make judicious guesses first and verify answers thereafter.

6. What is the length in inches of a diameter of a penny?

(a) $\frac{3}{8}$ (b) $\frac{1}{2}$ (c) $\frac{5}{8}$ (d) $\frac{3}{4}$ (e) $\frac{7}{8}$

7. Approximately how many pennies are there in 1 pound of pennies?

(a) 100 (b) 150 (c) 200 (d) 250 (e) 300

8. How many pennies must be stacked one on top of the other so that the height of the pile will be equal to the height of a quarter standing on edge?

    (a) 12   (b) 14   (c) 16   (d) 18   (e) 22

9. The length of a $1 bill is about how many times its width?

    (a) $\frac{5}{3}$ (b) $\frac{6}{3}$ (c) $\frac{7}{3}$ (d) $\frac{8}{3}$ (e) $\frac{9}{3}$

10. Which of the following segments has a length of 1 inch?

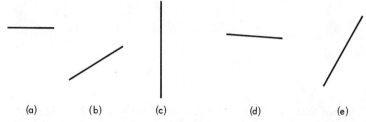

    (a)   (b)   (c)   (d)   (e)

11. Which of the following is the best estimate of the size of the given angle?

    (a) 45°
    (b) 30°
    (c) 55°
    (d) 20°
    (e) 50°

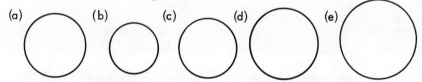

12. Which of the following shows the size of a nickel?

    (a)   (b)   (c)   (d)   (e)

13. Which of the following shows the width of a $1 bill?

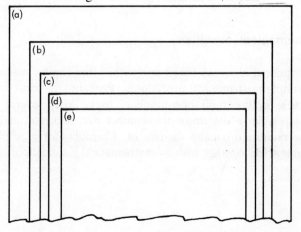

There are a number of geometric guessing activities that can lead to interesting discoveries. Most of these can be easily verified by the student through some simple construction or experimentation. Following are several representative items in this category.

14. Fold a piece of paper twice and cut off the folded corner as shown. When opened, the paper will have one hole.

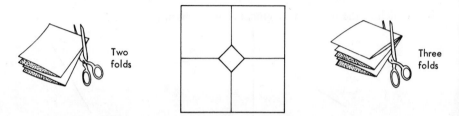

Suppose that the paper is folded three times and the folded corner is cut off. How many holes will there be in the paper? How many holes will the paper have if it is folded four times and then the folded corner is cut off?

15. Take a piece of paper and fold it down the middle as shown. If you then proceed to tear the paper as indicated, you will get three pieces. Suppose that the paper is folded a second time in the other direction before it is torn. How many pieces of paper will there be then?

How many pieces will there be if the paper is alternately folded three times before it is torn?

2.2 Mathemagical novelties

Many motivational ideas are based on "tricks" that can be justified through relatively simple mathematical procedures. Probably the most famous type of all of these is the "think-of-a-number" trick. After several operations the teacher is able to determine the number that a person ends up with, but *not* the number originally thought of. Consider, for example, the following simple trick together with its mathematical justification:

The teacher says:	*The teacher thinks:*
Think of a number.	n
Double the number.	$2n$
Add 7.	$2n + 7$
Subtract 1.	$2n + 6$
Divide by 2.	$n + 3$
Subtract the number you originally started with.	3 $(n + 3 - n = 3)$

At this point everyone in the class is thinking of the number 3. The teacher can then announce that this number is 3, or he can continue several additional operations to arrive at a special number, such as that day's date. This type of activity is very worthwhile to generate interest in mathematics, but it can very specifically be used to introduce a unit of work on variables as well as to show an application of the use of variables in algebra.

The following collection of mathematical tricks is illustrative of the type suitable for use in the classroom for purposes of motivation. In each case, the reader is invited to discover why the trick works as it does. Most of these explanations are within the grasp of average secondary school students, but even where they are not, it is still worthwhile to use the trick as a means of creating and maintaining interest.

1. *Instructions*   *Example*
   Write a three-digit number such that   872
       the hundred's digit is at least two more   − 278
       than the unit's digit.   ‾‾‾‾‾
   Reverse the digits and subtract.   594
   Reverse the digits again and add.   594
   The result will always be 1089.   + 495
                                                                ‾‾‾‾‾
                                                                  1089

2. *Instructions*   *Example*
   Write a two-digit number between 50   78
       and 100.   + 54
   Add 54.   ‾‾‾‾
   Cross out the hundred's digit, and   132
       add it to the remaining two-digit number.   ͳ32;   32 + 1 = 33
   Subtract your result from the number
       with which you started.   78
   The result will always be 45.   − 33
                                                                ‾‾‾‾
                                                                   45

Note: The final outcome depends upon the number added in the second step. Merely subtract this number from 99 to determine the outcome. Thus, if 84 were to be added, the final result would be 15.

In each of the following tricks, the participant must give the teacher certain information at a given time. Thereafter the teacher uses this information to determine some special number. Often such tricks are used to guess a person's age, as in the next item.

3. *Instructions*   *Example*

     Think of your age (or any other   14
        number greater than 9).

     Multiply by 10. $14 \times 10 = 140$

     From this number subtract the product   140
        of any one-digit number and 9. That is, $-\ 27$
        subtract some multiple of 9 from 9 $\overline{113}$
        through 81.

At this point the teacher asks the student to state the final outcome, 113. To determine the original number, cross out the unit's digit and add it to the remaining two-digit number. In the example shown, cross out the 3 and then add, to obtain $11 + 3 = 14$, the number started with.

4. *Instructions*   *Example*

     Think of a number.   21
     Multiply by 2. $21 \times 2 = 42$
     Add 5. $42 + 5 = 47$
     Multiply by 5. $47 \times 5 = 235$

At this point the student is asked to state the final outcome. To find the original number, merely remove the unit's digit and subtract 2 from the remaining two-digit number. In the example shown, cross out the 5 and then subtract 2 to obtain $23 - 2 = 21$.

5. *Instructions*   *Example*

     Think of your age.   15
     Multiply your age by 2. $15 \times\ 2 =\ \ 30$
     Add 10. $30 + 10 =\ \ 40$
     Multiply by 5. $40 \times\ \ 5 = 200$
     Add the number of people in your family. $200 +\ \ 4 = 204$
     Subtract 50. $204 - 50 = 154$

Now ask for the final result. The person's age will be represented by the hundred's and ten's digit, the number of people in the family by the unit's digit. For this example, 154 is interpreted to mean an age of 15 and a family size of 4.

6. *Instructions*	*Example*
Think of a number.	23
Multiply by 5.	$23 \times 5 = 115$
Add 6.	$115 + 6 = 121$
Multiply by 4.	$121 \times 4 = 484$
Add 9.	$484 + 9 = 493$
Multiply by 5.	$493 \times 5 = 2465$

Ask for the final result, and subtract 165 from this number. Drop the zeros in the unit's and ten's places, and the resulting number is the one started with. In the example shown we have $2465 - 165 = 2300$, and the original number thought of was 23.

Many tricks can appear to be quite dramatic through the use of extra showmanship. For example, the next trick begins with a book of matches and requires some burning. If it's too dramatic in this form, use candy instead of matches.

7. *Instructions*	*Example*
Hand a student a full book of 20 matches. Have him pick any number from 1 through 10 and remove that many matches from the book.	Assume that the student removes 8 matches.
Ask him to count the remaining matches and find the sum of the two digits in that number.	12 matches remain. The sum of the two digits in 12 is 3.
Have him remove that many more matches.	He removes 3 more matches.
Direct him to burn as many of the remaining matches as he wants, giving you the burnt matches that result.	Suppose that he chooses to burn 4 of the remaining matches. He gives back 4 burnt matches.
You immediately tell him how many matches are left in the book.	Subtract 4 from 9 to get 5, the number of matches left in the book.

Regardless of how many matches are removed by the student in the first step, there will be exactly 9 left at the end of the third step. So the teacher needs only to subtract the number of burnt matches returned from 9 to find how many remain in the book.

This trick will work as long as the student begins with a collection of 20 items. Twenty matches make a convenient book, but 20 candies will do as well. If the student starts with 20 candies and returns those that are left at the end, the teacher can "guess" how many he ate.

The following trick is an interesting one in that it involves two students.

8. Place a pencil and an eraser on the desk, and ask for two volunteers (we shall call them John and Mary).

Instructions	*.Example*
One of you take the pencil, and the other take the eraser. Don't tell me which one takes which object; I'll guess it!	Assume that John takes the pencil and Mary takes the eraser.
The one who took the pencil is assigned the number 7; the one who took the eraser has the number 9. (The teacher does not know who has which number at this stage.)	John begins with 7, and Mary begins with 9.
John, multiply your number by 2.	John: $7 \times 2 = 14$.
Mary, multiply your number by 3.	Mary: $9 \times 3 = 27$.
Find the sum of these products.	$14 + 27 = 41$.

The teacher now asks for this sum. If the sum is divisible by 3, then Mary took the pencil and John took the eraser. If the sum is not divisible by 3, then John took the pencil and Mary took the eraser. In the example shown, the sum of 41 is not divisible by 3, which indicates that John took the pencil.

As one final mathemagical trick, we consider an example that is appropriate for use in a secondary algebra class. The explanation for this one will be discussed in detail in that it depends upon the mathematics of finite differences and as such is an excellent application to motivate students in their study of algebra.

9. *Instructions*   *Example*

Think of a quadratic expression of
the type $ax^2 + bx + c$. $3x^2 - 5x + 2$

Substitute 0, 1, and 2 for x in that order,   For $x = 0$: 2
and give me the results. $x = 1$: 0
                                                                      $x = 2$: 4

To determine the original expression, the teacher proceeds to find first and second differences, as follows:

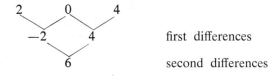

first differences

second differences

$(0 - 2 = -2; \quad 4 - 0 = 4)$
$[4 - (-2) = 6]$

The coefficient of the x^2 term is one half of the bottom number; $\frac{1}{2}$ of $6 = 3$. The coefficient of the x term is the first number of the middle row minus one half of the bottom number; $-2 \; -3 = -5$. The first number of the top row is the constant, 2. Thus, the original expression is $3x^2 \; -5x + 2$.

The explanation for these rules can be found by considering the general case, $f(x) = ax^2 + bx + c$. For this we have:

$$f(0) = c$$
$$f(1) = a + b + c$$
$$f(2) = 4a + 2b + c$$

Finding first and second differences we have the following:

$$c \searrow \quad a+b+c \searrow \quad 4a+2b+c$$
$$a+b \searrow \qquad 3a+b \nearrow$$
$$2a$$

From this array it is clear that one-half of the bottom number is a, the coefficient of x^2. The first number of the middle row minus one half of the bottom number gives $a + b - a = b$, the coefficient of x. The first number of the top row is c, the constant term.

Although the items presented throughout this chapter are included as suggested means for motivation, as well as simply for interest, it should be noted that many of them can be used to introduce topics of significant mathematical importance. This last item, for example, can lead into a discussion of finite differences and the procedure for finding an nth-degree polynomial if at least $(n + 1)$ data points are given.

To illustrate such an approach, let us consider the problem of determining the number of regions into which a circle can be divided by n chords. By experimentation we find the following:

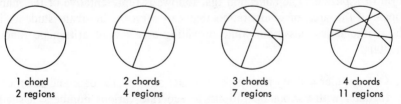

| 1 chord | 2 chords | 3 chords | 4 chords |
| 2 regions | 4 regions | 7 regions | 11 regions |

Collecting these data in a table and finding first and second differences produces the following information:

Number of chords, x	Number of regions, y	First differences	Second differences
1	2		
		2	
2	4		1
		3	
3	7		1
		4	
4	11		

Inasmuch as the second differences are constant, we assume that the formula relating the number of regions (y) as a function of the number of chords (x) is given by the second-degree function $y = ax^2 + bx + c$. To solve for the constants $a, b,$ and c we proceed to substitute the values given in the table and solve the three linear equations thus obtained.

$$\text{For } x = 1: \quad ax^2 + bx + c = a + b + c = 2$$
$$\text{For } x = 2: \quad ax^2 + bx + c = 4a + 2b + c = 4$$
$$\text{For } x = 3: \quad ax^2 + bx + c = 9a + 3b + c = 7$$

By subtracting the first equation from the second, and the second equation from the third, we obtain

$$3a + b = 2$$
$$5a + b = 3$$

Subtracting again gives $2a = 1$, or $a = \frac{1}{2}$. By substitution we find $b = \frac{1}{2}$ and $c = 1$. Thus, the number of regions, y, is given in terms of the number of chords, x, by the equation $y = \frac{1}{2}x^2 + \frac{1}{2}x + 1$. For $x = 4$, we find $y = 11$, as given in the table. Verify with a drawing that five chords will produce 16 regions.

For an alternative approach to this problem, see Experiment 1 of Section 4.5.

2.3 Computational curiosities

Many interesting computational curiosities are available to motivate instruction in mathematics. These range from items that the teacher can use for classroom demonstration to those that can be given to students in the form of a puzzle. The collection that follows is representative of the many topics in the area of computation that can serve to motivate students in the mathematics classroom while providing a review of arithmetic fundamentals.

1. One popular activity is to take the digits of a specific date and use these, together with various operations, to generate various numbers. Usually the additional stipulation is stated that the digits must be used in the order in which they appear in the date under consideration. As an example, consider the date 1776. Here are two ways to use the digits of this date to obtain the number 4:

$$17 - 7 - 6 = 4 \qquad (17 + 7) \div 6 = 4$$

An interesting project is to attempt to represent as many numbers as possible from 0 through 100 using these four digits in order. Depending upon the level of ability of the class, various operations may be permitted.

0: $1^7 - 7 + 6$
1: 1^{776}
$\left.\right\}$ with exponents

2: $1 + \sqrt{7 \times 7} - 6$
3: $\sqrt{1 + 7 + 7 - 6}$
$\left.\right\}$ with square roots

4: $|1 + (7 \div 7) - 6|$
5: $|1 - 7| - 7 + 6$
$\left.\right\}$ with absolute values

6: $1 \times (7 - 7)! \times 6$
7: $1 \times 7 \times (7 - 6)!$
$\left.\right\}$ with factorials

8: $1 + (7 \div 7) + 6$
9: $(1 + 7 + 7) - 6$
$\left.\right\}$ with fundamental operations only

This item can be used in a variety of ways. For example, it can be given as an assignment, it can be used as the basis for a race between chosen teams, it can form the basis for a bulletin board display, and so forth. It is most appropriate to use near the start of a new year, using the digits of the forthcoming year. For example, a class may be asked to represent the numbers from 1 through 10 using the digits 1975 in that order.

2. Closely related to the preceding item is one that requires as many numbers as possible to be represented using four 4's and any available operation. Here, for example, are representations for the numbers 0 through 10:

$$0: \quad 44 - 44$$
$$1: \quad 44 \div 44$$
$$2: \quad (4 \div 4) + (4 \div 4)$$
$$3: \quad (4 + 4 + 4) \div 4$$
$$4: \quad 4 + 4(4 - 4)$$
$$5: \quad 4 + 4^{(4 - 4)}$$
$$6: \quad \sqrt{4 \times 4} + 4 - \sqrt{4}$$
$$7: \quad 44 \div 4 - 4$$
$$8: \quad (4 \times 4) - 4 - 4$$
$$9: \quad 4 + 4 + (4 \div 4)$$
$$10: \quad (44 - 4) \div 4$$

3. Students can be motivated to review arithmetic in disguise through the use of secret messages that they can uncover by doing the appropriate problems. The message can be related to a particular event, such as "Season's Greetings" or "Happy Easter" or can merely say "No Homework," as in the following example.

DIRECTIONS

Work each problem. Then find the letter of the alphabet that corresponds to the answer. For example, the first problem is $126 \div 9 = 14$, and the fourteenth letter of the alphabet is N. After all problems are completed, the message is read down in the last column.

PROBLEM	NUMBER	LETTER
$126 \div 9$	14	N
75% of 20		
$\frac{2}{3} \times 12$		
$101 - 86$		
$\sqrt{169}$		
$5^2 \div 5$		
$(7 \times 5) - (4 \times 3)$		
$0.45 \div 0.03$		
2×3^2		
$(11 \times 11) \div 11$		

Depending upon the level of ability of the class, the problem can be made more or less difficult and can be designed to include extensive work with fractions, decimals, and per cents.

4. Finger computation is a topic that can generally be counted upon to develop interest in most classes. Multiplication by 9 on one's fingers is very interesting and can be accomplished as shown in the following diagrams.

    To multiply 3×9, bend down the third finger from the left. Then read the answer in groups of fingers on either side of the bent finger:

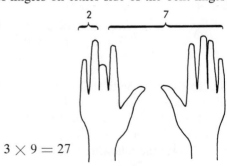

$$3 \times 9 = 27$$

Here are several additional examples:

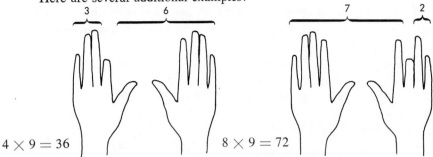

$4 \times 9 = 36$ $8 \times 9 = 72$

Fingers can be used to multiply a two-digit number by 9 provided that the unit's digit is greater than the ten's digit. To multiply 9×38, first put a space after the third finger from the left. Then bend the eighth finger from the left. Read the answer in terms of groups of fingers:

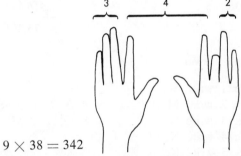

$9 \times 38 = 342$

5. "Magic numbers" can be used to fascinate students and motivate arithmetic review in a disguised form. For example, the magic number 15,873 works with 7. If 15,873 is multiplied by $7 \times n$, where n is a number from 1 through 9, the product consists of repeated n's. An interesting way to use this item in a class is to ask a student to give his favorite number from 1 through 9. Assume that he responds with 8. The teacher then tells him to multiply 15,873 by 56 (7×8); the product will be 888,888. Here are several additional examples:

$$\begin{array}{lll}
\begin{array}{r}15,873 \\ \times\ 14 \\ \hline 222,222\end{array} \ (7 \times 2) &
\begin{array}{r}15,873 \\ \times\ 28 \\ \hline 444,444\end{array} \ (7 \times 4) &
\begin{array}{r}15,873 \\ \times\ 49 \\ \hline 777,777\end{array} \ (7 \times 7)
\end{array}$$

There are other such numbers that can be used in a similar fashion. For example, the number 8,547 works with multiples of 13 as shown here:

$$\begin{array}{lll}
\begin{array}{r}8,547 \\ \times\ 26 \\ \hline 222,222\end{array} \ (13 \times 2) &
\begin{array}{r}8,547 \\ \times\ 39 \\ \hline 333,333\end{array} \ (13 \times 3) &
\begin{array}{r}8,547 \\ \times\ 65 \\ \hline 555,555\end{array} \ (13 \times 5)
\end{array}$$

6. The next item is another vehicle for presenting a secret message and is partially based upon an earlier item presented in Section 2.2. The following set of directions will produce the message "No Homework."

Directions	Example
Write any three-digit number so that the hundred's digit is at least two more than the unit's digit.	582
Reverse the digits and subtract.	$\begin{array}{r} -\ 285 \\ \hline 297 \end{array}$
Reverse the digits and add.	$\begin{array}{r} 297 \\ +\ 792 \\ \hline 1089 \end{array}$
Multiply by 10,000,000,000. Subtract 6,516,738,615.	$\begin{array}{r} 10,890,000,000 \\ -\ 6,516,738,615 \\ \hline 4,373,261,385 \end{array}$

Consider your answer.

Under each 1 write W.
Under each 2 write M.
Under each 3 write O.
Under each 4 write N.
Under each 5 write K.
Under each 6 write E.
Under each 7 write H.
Under each 8 write R.

4 3 7 3 2 6 1 3 8 5
N O H O M E W O R K

7. A difficult but interesting project is to use all nine digits from 1 through 9 to represent 100. Challenge a class to find as many different ways as possible to do this, and make a collection of these on a bulletin board. Here are several examples:

$$1 + 2 + 3 + 4 + 5 + 6 + 7 + (8 \times 9) = 100$$
$$123 - 45 - 67 + 89 = 100$$
$$1 + 2 + 3 - 4 + 5 + 6 + 78 + 9 = 100$$
$$(56 + 34 + 8 + 2) \times (9 - 7 - 1) = 100$$
$$62 + 38 + [(1 + 7) \times (9 - 5 - 4)] = 100$$

To make it more interesting the teacher may wish to restrict the use of the nine digits to their natural order, as in the first three examples above.

2.4 Geometric tidbits

Every teacher should have his own special collection of geometric tidbits—short little puzzles, problems, and curiosities in geometry to warm up the class, to gain attention, to involve, to challenge, to maintain interest, or simply to give a change of pace. The examples given here are but a few of the vast supply available in the literature or ready for the teacher to create. They are ideas for motivation as compared to more extensive classroom activities and experiments such as those found in later chapters. The particular use of these ideas in the mathematics class is left for the vivid imagination of the teacher. Answers are given at the end of the chapter.

1. Move just three dots to form an arrow pointing down instead of up.

2. Form four equilateral triangles with just six toothpicks.

3. How many pennies can you arrange such that each penny touches every other penny?

4. To mount a picture two thumbtacks are needed in any two corners. What is the least number of tacks needed to mount four pictures?

5. Rearrange three toothpicks to form a figure that consists of three squares of the same size.

6. Rearrange three toothpicks to form a figure that consists of five squares of the same size.

7. The digits 0 through 8 have been classified with the letters A, B, and C. How would you classify the digit 9?

A	1	4	7
B	2	5	
C	0	3	6	8

8. Here the letters A through H have been classified with the digits 1, 3, and 5. Discover the secret and classify the rest of the letters in the alphabet.

1	A E F H
3	C
5	B D G

9. These 18 matchsticks form a regular hexagon divided into three congruent regions. Move just four matchsticks to divide it into only two congruent regions.

10. A solid has this for both its top and front view. Draw its side view.

11. Take a single strip of paper and fold it into a regular pentagon.

12. Without lifting your pencil from the paper, try to draw four connected lines that pass through all nine points. Remember, the lines must be straight.

```
•  •  •
•  •  •
•  •  •
```

A whole host of interesting little geometric curiosities involve counting. Here are some examples.

13. How many triangles are in this figure?

14. How many squares are in this figure?

15. How many rectangles are in this figure?

16. Count all the 1 × 1 × 1, 2 × 2 × 2, and 3 × 3 × 3 cubes in this figure.

Another type of geometric tidbit that can be considerably more challenging requires skill in space perception. The following illustrate problems of this type. In each case, the student should guess at his answers first and then construct and use models to verify them.

17. List all the possible paths along the edges of cube that pass through all eight vertices just once in going from vertex *A* to vertex *G*.

18. A tetrahedron and a cube are to be painted but no two adjacent faces on either can be colored the same. What is the least number of colors needed for each?

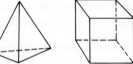

19. Describe the shape of the surface formed by spinning the cube through the axis shown.

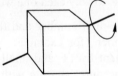

20. Imagine a tetrahedron cut straight through the midpoints of three sides as shown. Describe the shape of each intersection.

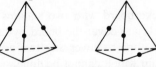

21. Imagine a cube cut straight through the midpoints of three sides as shown. Describe the shape of each intersection.

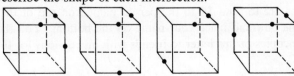

22. Describe the shape of a single solid that will fit snugly through each of these openings.

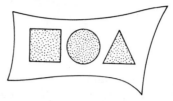

Exercises

1. Represent the numbers from 1 through 20 using the digits of the current year in the order in which they appear. Any familiar mathematical operation may be used.

2. Extend the list of fours on page 35 to represent the numbers from 11 through 20.

3. About how many pennies would have to be piled one on top of another to reach the ceiling of a room that is 8 feet high?

4. To the nearest thousand, how many pennies are there in 1 mile of pennies placed next to each other with their edges touching?

5. How long would it take to spend 1 million dollars at the rate of $100 every minute?

6. Section 2.2 lists a number of mathematical tricks. Provide a mathematical explanation as to why each one works.

7. Write down any three-digit number and then repeat the same digits to form a six-digit number (for example: 345,345). Now divide by 7, then divide the quotient by 11, and finally divide this quotient by 13. Repeat the same procedure, starting with another three-digit number. State your discovery, and then provide a mathematical explanation as to why this trick works as it does.

8. Consider a checkerboard with two squares missing at diagonally opposite corners. Try to cover the board by use of a set of dominos that are only large enough to cover any two adjacent squares. Show your solution or explain why it cannot be done.

9. You receive a chain letter with a list of five names on it. Your instructions are to send $1 to the person named on top of the list, cross his name out, and place your name on the bottom. You then are to send five copies of this letter to friends with instructions for them to proceed

in the same manner. Assume that this process continues and that no one breaks the chain; how much money should you ultimately receive?

10. Repeat Exercise 9 for a chain letter with 10 names.

Activities

1. Prepare a multiple-choice set of questions, similar to those found on page 26, that requires one to guess about familiar objects.

2. Examine at least two junior high school textbooks and report on the use made of motivational techniques in each. In particular, report on any unusual approaches to motivation that are used.

3. Repeat Activity 2 for two senior high school mathematics books.

4. Repeat Activity 2 for a textbook that is specifically designed for slow learners in mathematics.

5. Read and report on Chapter 6, "The Art of Generating Interest," by Herman Rosenberg, in the Thirty-third Yearbook of the National Council of Teachers of Mathematics, *The Teaching of Secondary School Mathematics.*

6. Repeat Activity 5 for Chapter 1, "Skills," by Max A. Sobel. In particular, consider suggested techniques for motivation in the area of skill development.

7. Develop a one-period lesson plan for a junior high school mathematics class that clearly indicates procedures to be used to motivate the lesson.

8. Repeat Activity 7 for a senior high school mathematics class.

9. Prepare a collection of at least 10 "geometric tidbits" similar to those given in Section 2.4.

10. Prepare several secret-message charts such as the one shown on page 36. Have one of these feature computation with fractions, another computation with decimals.

References and selected readings

ABBOTT, EDWIN A. *Flatland.* New York: Barnes & Noble, Publishers, 1963.

BERGAMINI, DAVID, and the Editors of *Life. Mathematics.* Chicago: Time-Life Science Library, 1963.

DUDLEY, H. E. *Amusements in Mathematics.* New York: Dover Publications, Inc., 1958.

EVES, HOWARD. *In Mathematical Circles.* Boston: Prindle, Weber & Schmidt, Inc., 1969.

FLETCHER, T. J. *Some Lessons in Mathematics.* New York: Cambridge University Press, 1964.

JACOBS, HAROLD R. *Mathematics, a Human Endeavor.* San Francisco: W. H. Freeman and Company, Publishers, 1970.

JOHNSON, DONOVAN A., and G. R. RISING. *Guidelines for Teaching Mathematics.* Belmont, Calif.: Wadsworth Publishing Company, Inc., 1972, Chap. 9, "Developing Positive Attitudes and Creativity Through Enrichment."

MADACHY, JOSEPH S. *Mathematics on Vacation.* New York: Charles Scribner's Sons, 1966.

Mathematics Teacher. "Garbage Collection, Sunday Strolls, and Soldering Problems" (Apr. 1972), p. 307.

———. "Mathematical Misteaks" (Feb. 1971), p. 109.

———. "Space Filling in Two Dimensions" (Nov. 1971), p. 587.

———. "A Tiny Treasury of Tessellations" (Feb. 1968), p. 114.

MIRA, JULIO A. *Mathematical Teasers.* New York: Barnes & Noble, Publishers, 1970.

PECK, LYMAN. *Secret Codes, Remainder Arithmetic, and Matrices.* Washington, D.C.: National Council of Teachers of Mathematics, 1961.

SOBEL, MAX A., EVAN M. MALETSKY, and T. HILL. *Essentials of Mathematics, Books 1, 2, 3, 4.* Lexington, Mass.: Ginn and Company, 1969, 1970, 1973.

STEINHAUS, HUGO. *Mathematical Snapshots.* New York: Oxford University Press, Inc., 1969.

Answers to section 2.4: geometric tidbits

1.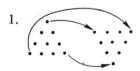

2. Form a tetrahedron.
3. 5
4. 3

5. 　　　6.

7. C (formed with curves only)
8. 1—I K L M N T V W X Y Z　　3—O S　　5—J P Q R U

9.

10. Other answers are possible.

11. Tie an overhand knot with the paper strip.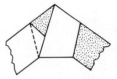

12.
　Start

13. 10
14. 14
15. 40
16. 36
17. ABCDHEFG　ABFEHDCG　ADCBFEHG　ADHEFBCG
　　AEHDCBFG　AEFBCDHG
18. 4; 3
19.

20. Equilateral triangle; square
21. Equilateral triangle; rectangle; regular hexagon; pentagon
22. Start with a cylinder with a square cross section. Then taper from the top.

CHAPTER THREE

RECREATIONAL ACTIVITIES

Mathematical recreations can serve as a very effective means of motivation at almost all levels of instruction and for students of varying levels of ability. The supply of such items is almost inexhaustible and can be used in many ways. Some activities can be used as an integral part of the daily lesson, others can be used to promote discovery in laboratory settings, and still others are worth introducing just for fun. The regular use of mathematical recreations throughout the year can very well serve as a means of convincing students that mathematics can be quite exciting, with hopeful transfer to normal activities.

Within this chapter are presented a variety of mathematical games, puzzles, and enrichment topics. For each of these categories, only a relatively small number of illustrative examples are given. Hopefully the reader will be stimulated to search the literature and follow the suggested bibliographical leads to begin a collection of many more of these recreational activities.

3.1 Mathematical games

Mathematical games can be used effectively for a variety of purposes. They can be used solely as recreational devices to motivate a class and generate interest. As such, they supply a good source of material for use during the last few minutes of a period, the day before a holiday, and in similar circumstances. Their use as part of the program for a mathematics club can almost always be counted upon to interest student members.

In addition to purely recreational aspects of games, other objectives can be attained through their use. Many mathematical games can be used to lead students to formulation and testing of hypotheses as they strive to discover a winning strategy. The development of such modes of thinking has long been recognized as a worthwhile outcome of the teaching of mathematics.

Finally, mathematical games can be used as an effective way to develop certain basic concepts and skills. Arithmetic and geometric skills, as well as ability to visualize in two and three dimensions, are just a few of the many mathematical items that can be approached through the use of appropriate games.

The following collection is only a representative sample of suitable games for the mathematical classroom. The reader is referred to the bibliography on page 82 for further sources of information on this topic.

The game of 50

The Game of 50 is a game designed for two individuals. An effective way to introduce the game is for the teacher to announce that he is the world champion Game of 50 player, and challenge a student to try to win. Thereafter, students can play against one another.

RULES FOR PLAYING

The game is played using the numbers 1, 2, 3, 4, 5, and 6. The two players alternate in selecting numbers, and the first to reach 50 wins. As each new number is selected, it is added to the sum of the previously selected numbers. For example, if the student goes first and selects 3, the teacher might then select 6, to give a sum of 9. If the student then selects 5, the total is 14, and it becomes the teacher's turn to go. The game continues in this manner until one player becomes the winner by reaching 50.

STRATEGY

Analysis of the game shows that you can always reach 50 if you first reach 43. (Regardless of what number your opponent selects, you can choose a number to obtain 50.) Working backward, you can reach 43 if you can get to 36. Continuing in this way, the following "winning numbers" are obtained:

$$1, \ 8, \ 15, \ 22, \ 29, \ 36, \ 43, \ 50$$

Thus the strategy for winning is to go first and begin with 1. Thereafter, select the complement of your opponent's number relative to 7. That is, if he picks 4, you choose 3, if he picks 2 you choose 5, and so forth. If your opponent goes first, work to one of the winning numbers as soon as you can.

EXTENSIONS

Many variations of this game are possible. A similar game is played by using a set of 16 cards consisting of the four aces, four 2's, four 3's, and four 4's.

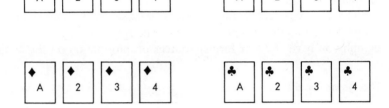

Players alternate selecting one card at a time from the pile of 16 cards, without replacements. As before, cumulative sums are kept. The winner is the first person to select a card that brings the total to exactly 22, or forces his opponent to go over 22. The set of numbers here is {2, 7, 12, 17, 22}. The strategy for winning is to go first and begin with 2. Thereafter, select the complement of your opponent's number relative to 5. That is, if he picks 4, you choose 1, and so forth. However, this is not a foolproof strategy because the number of cards is limited. Suppose that your opponent repeatedly chooses 3. This would force you to repeatedly choose 2 and you would run out of 2's prior to reaching the objective number of 22. Can you consider alternative strategies for such a situation?

The game of sprouts

The game of Sprouts is a game for two that is most effectively presented by dividing the class into pairs to play against one another. The one who wins two out of three games can then be declared the winner, and the winners paired against one another again until a class champion emerges.

RULES FOR PLAYING

The game starts with two points, labeled *A* and *B* in the figure and called *spots*. Each player takes a turn drawing an arc from one spot to another, or to the same spot. He then places a new spot on his arc. For example, here are two possible moves that the first player might make:

The two basic rules of play are that no arc may cross itself or pass through another arc or spot, and that no spot may have more than three arcs from that point. The winner is the last person who is able to draw an arc. Here is an example of a game played; a circle around a point indicates that there are three arcs at the point and the point is thus no longer in play:

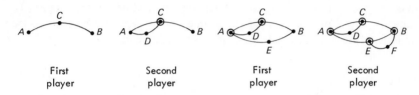

First	Second	First	Second
player	player	player	player

In the game illustrated, the second player wins. The first player cannot draw an arc from D to F because it would pass through another arc. Also, he cannot draw arcs from D to itself (or from E to itself) because then there would be four arcs at the point.

STRATEGY

Note that each spot becomes a dead spot once it is used three times as the endpoint of an arc. Therefore, the game begins with six possible "lives." The first player uses up two lives, but adds one when he adds a spot to his arc. Therefore, after the first play the game has five lives left. In a similar manner, there are four lives left after the second move, and only one life left after the fifth move. With only one arc available, the game must be at an end. Therefore, the game has a maximum of five possible moves, a small enough number for a student to be able to draw the set of all possible games.

Here is an example of a game that utilizes the maximum number of moves, resulting in a win for the first player:

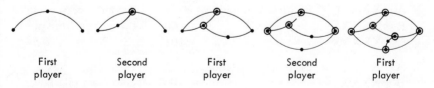

First	Second	First	Second	First
player	player	player	player	player

EXTENSIONS

The game can be played using any number of spots initially. Consider playing the game using three spots, in which case there will be a maximum of eight possible moves. The maximum number of moves possible in four-spot sprouts is 11.

For further discussion and extensions, see Martin Gardner's column, "Mathematics Games," *Scientific American,* July 1967, pp. 112–115.

The game of Nim

Although Nim may be used purely as recreation at any time, it is best presented after the student has some knowledge of the binary system of numeration, thus providing an interesting application of that topic. The game is played by two students at a time and is probably best introduced by having the teacher challenge a student to a match.

RULES FOR PLAYING

Many versions of the game can be played. Probably the easiest one to start with consists of three piles of chips. Pile A contains 3 chips, pile B contains 4 chips, and pile C contains 5 chips. At a particular player's turn he selects one pile of chips and removes as many chips as he wishes from that one pile. He must, however, remove at least one chip. Players alternate, and the player who picks up the last chip on the board is the winner.

STRATEGY

Students may develop informal strategies for winning. The formal winning strategy is quite complex and consists of writing the number of chips in each pile in binary notation. In order to win, one must be certain that the sum of the digits in each binary place is even after his move. To illustrate this principle, here is an example of a game where the first player wins:

1.	A	XXX	In binary notation:	11	First player takes two
	B	XXXX		100	chips from pile A,
	C	XXXXX		101	as shown next.
2.	A	X	In binary notation:	1	Second player takes
	B	XXXX		100	three chips from pile C.
	C	XXXXX		101	
3.	A	X	In binary notation:	1	First player takes one
	B	XXXX		100	chip from pile B.
	C	XX		10	
4.	A	X	In binary notation:	1	Second player takes
	B	XXX		11	both chips in pile C.
	C	XX		10	

5. A   X   In binary notation: 1   First player takes two
   B   XXX   11   chips from pile B.

6. A   X   In binary notation: 1   Second player takes
   B   X   1   one of the chips, and
                                                   the first player wins on
                                                   the seventh move by
                                                   taking the remaining
                                                   chip.

Note in the above game that the first player always made a move that left the second player with an array of chips whose number in binary notation had a even number of 1's in each column.

EXTENSIONS

The game can also be played using the rule that the one who is forced to take the last chip is the loser. It can also be played with an indefinite number of chips in each of the original three piles, although the strategy for winning remains the same.

For further reading about this game, see the following:

BALL, W. W. R. *Mathematical Recreations and Essays.* New York: Macmillan Publishing Co., Inc., 1960, pp. 36–40.

KRAITCHIK, MAURICE. *Mathematical Recreations.* New York: Dover Publications, Inc., 1953, pp. 86–88.

The game of Tac Tix

Tac Tix, a game for two, is a variation of the game of Nim, which was invented by Piet Hein of Denmark in recent years. It is an interesting one to present in that it has not yet been completely analyzed.

RULES FOR PLAYING

Arrange 16 coins or chips, as in the figure. These are numbered here for ease of reference.

Players alternate removing any number of chips from any single row or column. However, as an additional constraint, only adjacent chips may be removed. For example, if player A removes chips 14 and 15 on his first move, player B may not take 13 and 16 in one move. The player who is forced to take the last chip is the loser.

STRATEGY

It is necessary to play this game so that the one who takes the last chip loses because otherwise the first player would always win. On a 3 x 3 board, using nine counters and following the rules stated above, the first player can win by taking the center chip, or a corner chip, or all of a central row or column. There is no known strategy for winning the game with 16 chips as described above.

EXTENSIONS

For further reading on this game see *Scientific American Book of Mathematical Puzzles and Diversions,* Martin Gardner, ed. (New York: Simon and Schuster, 1964, pp. 157–160). In this account Gardner suggests that one can best gain an introduction to this game by solving specific problems. Thus, it seems worthwhile to present Tac Tix problems for students to solve before actually playing the game with an opponent. Gardner presents these two problems suggested by the inventor, Piet Hein. In each case you are to find a move that will guarantee a win.

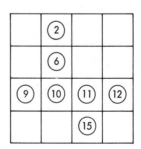

The game of Hit-and-Run

Hit-and-Run is another game for two that requires some careful thinking on the part of the players. It is most conveniently played on graph paper, and is similar to a number of commercial games currently on the market.

RULES FOR PLAYING

The game starts with a square grid. Players alternate coloring one line segment at a time in an effort to build a bridge from one side of the grid to the other. For example, the figure shows a winning path for player A that goes from his chosen side of the board to the opposite side. The subscripts indicate the order in which the moves were made. (Note that opponent's paths may cross one another.)

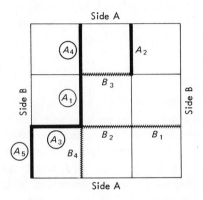

STRATEGY

For a 2 x 2 gird, the first player can always win by following the indicated moves.

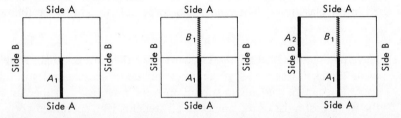

At this point A has two possible paths to complete, and therefore must win regardless of where B moves.

For a 3 x 3 grid, the first player has an advantage if his first move is in one of the positions shown.

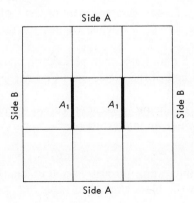

EXTENSIONS
    Play the game on a 4 x 4 grid. At the moment it is not known whether
there is a winning strategy for this size game nor whether there is any
advantage to going first. You can read more about this game, as well as
suggested extensions, in Martin Gardner's column, "Mathematical Games,"
Scientific America, July 1969, pp. 117–118.

3.2 Puzzle pastimes

Most students of mathematics enjoy working with puzzles. Although these
are most often recreational in nature, nevertheless there are many other
worthwhile outcomes that accrue from the use of suitable puzzles and problems. Appropriate recreational items can be found to stimulate intellectual
curiosity, to develop abilities in space perception, to promote discovery,
and to develop modes of thinking. The major purpose for their use in the
classroom, however, is to stimulate interest in the further study of mathematics.
    There are a variety of effective ways that mathematical recreations can
be used in the classroom. The reader should be able to add to the following list of suggested uses:

(a) Have one section of the bulletin board or chalkboard entitled "Puzzle
     of the Week." Each week place a new puzzle or set of puzzles in this
     spot. Encourage students to submit answers in writing. At the end of
     the week post a list of the names of all students who submitted correct
     solutions. Possibly award a prize for the first correct answer (such as
     an excuse from homework for one night), or a prize to the student
     who has the greatest number of correct solutions within each marking
     period.
(b) Place a collection of puzzles on 4- by 6-inch index cards. Any student
     who completes a classroom assignment or test early is allowed to come
     up and select a card to use while waiting for the rest of the class to
     finish their work.
(c) Devote the last 5 minutes of each period to a mathematical recreation.
(d) Devote the last half of the period each Friday to mathematical recrea-
     tions.
(e) Use the last period before a vacation period for recreational activities.
(f) On occasion include an interesting puzzle as part of the regular assign-
     ment.

    The supply of interesting puzzles is almost an endless one, and the
reader is referred to the bibliography on page 82 for additional sources

of such items. The following list is only representative of the type of puzzle that appears to be of interest to secondary school mathematics students. Answers to all these puzzles may be found at the end of the chapter.

1. How can you cook an egg for exactly 15 minutes, if all you have is a 7-minute hourglass and an 11-minute hourglass?
2. How can a 24-gallon can of water be divided evenly among three men with unmarked cans whose capacities are 5, 11, and 13 gallons?
3. Nine coins are in a bag. They all look alike, but one is counterfeit. It weighs less than the others. Use a balance scale and find the fake coin in exactly two weighings.
4. There are 12 coins, of which one is counterfeit, weighing less than the others. Use a balance scale and find the fake coin in exactly three weighings.
5. It is decided that whenever a certain club meets, everyone must shake hands with everyone else. If only two people meet, then only one handshake is necessary. If three people meet, then three handshakes are necessary.
   (a) How many handshakes are necessary if four people meet?
   (b) How many handshakes are necessary if five people meet?
6. A man went into a store and purchased a pair of shoes worth $5, paying with a $20 bill for his purchase. The merchants, unable to make change, asked the grocer next door to make change. He then gave the customer his shoes and the $15 change. After the customer had left, the grocer discovered that the $20 bill was counterfeit and demanded that the shoestore owner make good for it. The shoestore owner did so, and turned the counterfeit bill over to the FBI. How much did he lose by this transaction?

These first examples represent one type of numerical puzzle that can be found. Another type of puzzle deals with arrangements, usually with coins, cards, or toothpicks. These can be a bit more involved, but because they can be readily tried with actual objects, they can be lots of fun, too.

7. Use the digits 1, 2, 3, 4, 5, 6, 7, 8, and 9. Place exactly one of these in each position in the figure so that the sum along each side of the "magic" triangle is 20. As an extension, use the same digits and construct a "magic" triangle with the smallest and the largest possible sum along each side.

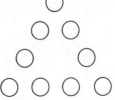

8. Use the digits 2, 3, 4, 5, 6, 7, 8, 9, and 10. Place exactly one of these in each position in the figure so that the sum for each row, column, and diagonal is 18.

9. Arrange the numerals 1 through 8 in the figure so that no two consecutive integers touch at a side or on a corner.

Examples of arrangements that are not permitted:

10. Arrange eight coins as in this figure:

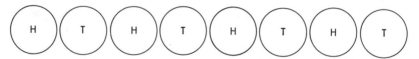

Moving two adjacent coins at a time, try to obtain the following arrangement:

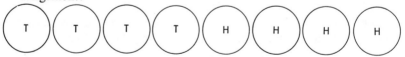

11. Interchange the penny and the nickel in the figure by sliding the coins from square to adjacent square, one at a time:

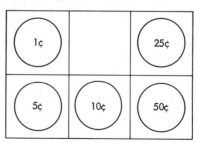

12. Draw a square, number eight pieces of paper, and arrange them as shown in the figure on the left. By sliding only one piece at a time into an open square, arrange the pieces as shown in the figure on the right:

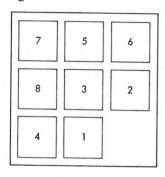 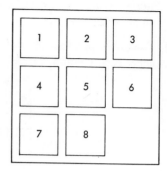

13. Arrange five coins as in this figure:

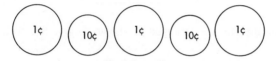

Try to obtain the following arrangement by moving two adjacent coins at a time, but each pair of coins moved must consist of a penny and a dime and must not be interchanged during the move:

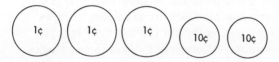

14. Place two pennies and two dimes on a set of five squares as in the figure:

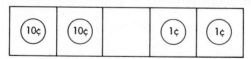

The object of the game is to interchange the positions of the pennies and the dimes. The rules are that dimes may be moved only to the right, and pennies may be moved only to the left. Coins may be jumped (without removal), but only one square at a time, as in checkers. The interchange can be made in a minimum of 8 moves. As an extension consider a set of seven squares, and use three pennies and three dimes. Follow the same rules and attempt to make the interchange in 15 moves.

15. Arrange three piles of toothpicks, chips, or other similar objects so as to have 6 items in pile A, 7 items in pile B, and 11 items in pile C. In exactly three moves you are to attempt to obtain 8 items in each pile. The rules for movement are that you may only move to a pile as many items as are already there, and all items moved must come from a single other pile.

16. Place five pennies on five squares so that no two pennies are in the same row, column, or along a diagonal:

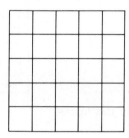

17. Eleven toothpicks are arranged as shown to give five triangles:

   (a) Remove one toothpick to show four triangles.
   (b) Remove two toothpicks to show four triangles.
   (c) Remove two toothpicks to show three triangles.
   (d) Remove three toothpicks to show three triangles.

18. A coin is in a "cup" formed by four matchsticks. Try to get the coin out of the cup by moving only two matchsticks to form a congruent "cup" but in a new position:

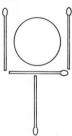

Additional puzzles such as the last two can be found in Section 2.4, Geometric Tidbits. Some geometric puzzles can lead to interesting generalizations, as illustrated by these on paths and networks.

19. Here are floor plans for two houses. In each case try to get from door *A* to door *B* by walking through each room once and only once.

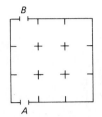 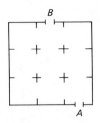

20. Can you walk through every door in this house exactly once, entering through the first and leaving through the last?

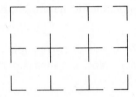

21. Try to walk through the maze. Start at the top, walk through each path exactly once, and come out at *B*. Then try to walk through each path once and only once and come out at *A* or *C*.

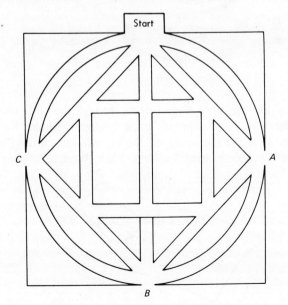

22. Three houses are located in the figure at *A, B,* and *C.* There are also three gates, numbered 1, 2, and 3. Draw a path from house *A* to gate 1, from house *B* to gate 2, and from house *C* to gate 3 so that no path crosses another.

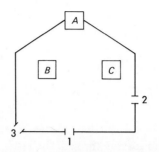

Still other puzzles of a geometric nature simply require great amounts of patience and ingenuity.

23. Loosely tie together the hands of two people as shown, with one string looped around the other. Now get them apart without untying or cutting the string. It can be done!

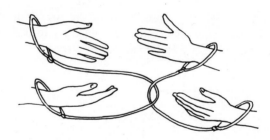

24. Cut a tag like this one. Pass a string through the hole, under the narrow center strip, out again, and back through the hole. Knot the ends. The trick is to get the string off the tag without cutting the string, or tearing the tag, or passing the string back through the hole.

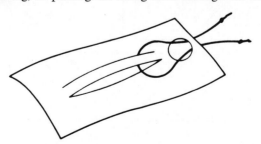

25. Find the pattern in this array, and do not give up too soon.

$$\circ \; \circ \; \triangle \; \circ \; \square \; \square \; \triangle$$
$$\square \; \square \; \square \; \circ \; \circ \; \circ \; \circ$$
$$\square \; \circ \; \circ \; \circ \; \triangle \; \circ \; \square$$
$$\circ \; \triangle \; \square \; \square \; \circ \; \triangle \; \square$$
$$\triangle \; \circ \; \circ \; \square \; \square \; \circ \; \circ$$
$$\circ \; \circ \; \square \; \square \; \circ \; \triangle \; \circ$$

Quite a different type of puzzle that can be both amusing and useful in the mathematics classroom deals with letter and word arrangements and codes. Some are straightforward and reasonably simple, whereas others can tax the most capable. Students enjoy making as well as solving these problems.

26. Each of the following will form a single familiar mathematical word when rearranged. How many can you decipher?

    (a) CAN IT FOR   (b) LUMTY LIP   (c) MAD LICE
    (d) CART TUBS   (e) ME RUN B   (f) I SON VIDI

27. Can you unscramble the names of these mathematicians?

    (a) A SCALP   (b) SEED CARTS   (c) UL DICE
    (d) SOAP THY RAG   (e) NOT NEW   (f) MAIDS CHEER

28. Can you break these alphametics?

Addition:	ONE	Subtraction:	FIVE
	TWO		FOUR
	FIVE		ONE
	EIGHT		

29. A student at college sent the following message to his father:

$$\begin{array}{r} \$\,W\,I\,.\,R\,E \\ M\,O\,.\,R\,E \\ \hline \$\,M\,O\,N\,.\,E\,Y \end{array}$$

If each letter represents a unique number, how much should his dad send?

30. More than 60 words, phrases, and names associated with geometry are hidden in this array. They are spelled horizontally, vertically, diagonally, and both forward and backward. How many can you find?

```
R T L U H F A C E I L C N A I D E M E Y P
A O D O O E T I N T E R C E P T A E R A O
L F T O G W E A I Q G J E B D L O R E Z L
I S R C V I L X S R S N A W T R X O H S Y
M P I E E Z C R A P I K M I O L O E P N G
I I A M C S R D X L V P T G R A P H S O O
S N N E A Q I U I A E U L P I U T T C I N
E D G R P U C B S N D N A M G Q T N D T D
Q U L T S A T N R E G R G E I E R A E C F
U C E X N R N E O A A O I T N D A E F E Z
I T M E O E Q G Q L L R N S H C P R I S M
L I P C Y R M D L O C U S O L G E O N C A
A O O L R O P E N T E U C L I D Z G I I E
T N L S T O L S L O P E R I I M O A T N B
E A K I E T J X I P K V I D D H I H I O T
R U N W M T V O L U M E B J A N D T O C D
A A R G O I Y M I U O O E I R X E Y N E F
L E C F E S T R V K I L C Z O B A P G N D
D G Q U G N S H H E X A G O N Y A R R A Y
P O I N T C T T J P A N H W B I E S G E C
A R B M D E D U C T I O N T H E O R E M P
```

Many of these puzzle pastimes lend themselves to a series of problems or activities that can be presented to the student in the form of laboratory worksheets or activity cards. Here is one example. Many others can be found in Chapter 4.

Activity card 1

Start with the A on top. Move down to the left or right one letter at a time. The path shown spells out the word ALGEBRA.

How many different paths are possible? Do they all spell ALGEBRA?

```
        A
       L L
      G G G
     E E E E
    B B B B B
   R R R R R R
  A A A A A A A
```

Ans. All 64 possible paths spell ALGEBRA.

Activity card 2

Start with an A on top. Move down
to the left or right, one letter at a
time. How many different ways can
you spell out the word ALGEBRA?

A A A A A A A   Ans. 64
L L L L L L
G G G G G
E E E E
B B B
R R
A

Activity card 3

Start with A and move down, or to
the right, or diagonally down and to
the right, one letter at a time. How
many ways can you spell out the
word ALGEBRA?

A L G E B R A   Ans. 127
L L-G E B R A
G G G E B R A
E E E E B-R A
B B B B B R A
R R R R R R A
A A A A A A A

3.3 Mathematics in familiar games

In addition to the recreational aspects of games, many games can be used
effectively in the classroom to motivate the study of specific mathematical
topics. In this section some mathematical applications are illustrated for
the familiar games of tic-tac-toe, checkers, chess, pool, poker, and roulette.

Tic-tac-toe

A careful study of two- and three-dimensional tic-tac-toe can produce
some interesting applications to counting and coordinates that can be both
useful and motivating in the mathematics classroom. Here are some suggested activities.

Set up the grid on a pair of axes, and play the game with the class using
ordered pairs of numbers to locate moves.

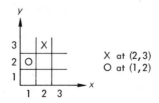

X at (2,3)
O at (1,2)

Present partially completed games and have the students suggest subsequent moves. For example, if these moves have been made, where should O move next?

O's at (1, 3), (2, 2), (1, 1)
X's at (3, 3), (1, 2), (3, 2)

Together there are eight possible winning arrangements in the game of tic-tac-toe. Have students give the number pair needed to make the winning arrangements.

(3, 1), (3, 3), __?__
(1, 3), (2, 2), __?__
(2, 2), (3, 2), __?__

Students can then practice giving the correct equations for different possible winning arrangements.

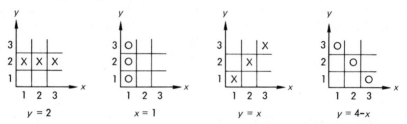

Applications from three-dimensional tic-tac-toe follow in much the same way only at a decidedly higher level of difficulty. Some suggested classroom activities are described here.

Set up the grid on three mutually perpendicular axes and practice locating positions using number triples in the form (x, y, z). Have students de-

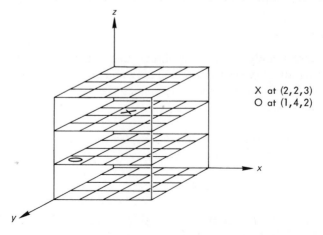

X at (2,2,3)
O at (1,4,2)

scribe some four-in-a-row wins. It is a nontrivial exercise simply to count the 76 possible winning arrangements on a 4 × 4 × 4 grid.

Students can give the num-   (2, 3, 1), (2, 3, 2), (2, 3, 3), ?
ber triple needed to com-   (1, 1, 1), (2, 2, 2), (3, 3, 3), ?
plete a winning arrangement   (4, 4, 4), (4, 3, 3), (4, 2, 2), ?
such as in these examples.

An interesting question would be to
find the seven winning arrangements   ? ,   ? , (3, 3, 3),   ?
containing a given position such as
(3, 3, 3).

If the idea of counting possible winning arrangements is extended to each of the 64 positions on the grid, the player would have a good idea as to the best moves to make initially in playing the game.

Pool

Some interesting applications of geometry can be found in the game of pool. For example, consider these properties and their possible use when teaching about angles.

The angle of rebound off a cushion is the same as the angle of approach.

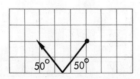

The angles of rebound off opposite cushions are the same. The new direction of the ball is parallel to its initial direction because the alternate interior angles are congruent.

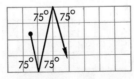

The angles of rebound off adjacent cushions are complementary. Here again the new direction of the ball is parallel to its initial direction.

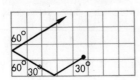

A variety of simple exercises illustrating these properties can be assigned using graph paper to establish the shape and dimension of the table and to facilitate the easy construction of congruent angles. For example, describe the path of the ball when shot in the direction shown.

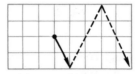

Checkers and chess

A variety of novel applications of checker and chess boards can attract attention in the mathematics class. An introduction or review of coordinate geometry can take on a new twist when presented through the games of checkers and chess.

As an example, what sequence of jumps can be made against black in this position on a checker board? Move from $(2, 4)$ to $(4, 6)$ to $(2, 8)$, capturing men at $(3, 5)$, and $(3, 7)$.

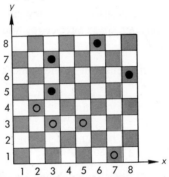

In chess, squares protected by various pieces can be given in coordinate form.

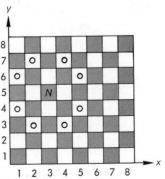

The knight at $(3, 5)$, protects squares
$(1, 4), (1, 6), (2, 3), (2, 7), (4, 3), (4, 7), (5, 4), (5, 6)$

A bishop at $(7, 4)$ would protect squares

$$(4, 1), (5, 2), (6, 3), (8, 5) \quad \text{and} \quad (3, 8), (4, 7), (5, 6), (6, 5), (8, 3)$$

The concept of an equation as a set of numbered pairs satisfying a given condition can be reinforced by expressing the coverage of various chess pieces.

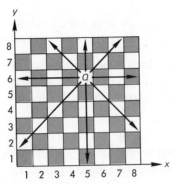

The queen at $(5, 6)$ protects all positions on these four lines:

$$x = 5, \quad y = 6, \quad y = x + 1, \quad \text{and} \quad y = 11 - x$$

A rook at $(2, 4)$ would protect all positions on these lines:

$$x = 2 \quad \text{and} \quad y = 4$$

Pascal's triangle can be developed through the idea of a random walk by a checker on a board. Start with the checker in square $(4, 7)$, and repeatedly toss a coin. If the coin falls heads, move down to the left and if it falls tails, move down to the right. The path of the checker can be

thought of as a random walk. The checker moves one square at a time diagonally down to either the left or the right. By counting the number of different ways it can move to various positions on successive rows, the numbers in Pascal's triangle are generated. These numbers can then be readily applied to the binomial expansion. The numbers in the third row represent the coefficients in the expansion of $(H + T)^3$.

$$(H + T)^3 = 1H^3 + 3H^2T + 3HT^2 + 1T^3$$

Each term also identifies the different random walks to that position.

$$\begin{array}{llll} \text{HHH} & \text{HHT} & \text{HTT} & \text{TTT} \\ & \text{HTH} & \text{THT} & \\ & \text{THH} & \text{TTH} & \end{array}$$

The concept of symmetry in geometry can be explained through the use of checkers and a board. Here are several suggestions.

> Locate a vertical, horizontal, or diagonal line on the board. Place a checker on one side of the line and have a student place another symmetric to it about the given line to illustrate line symmetry.

> Locate a square on the board. Then each time you place a checker on the board, have a student place another symmetric to it from the original square to illustrate point symmetry.

There are, of course, a great many interesting chess puzzles as well.

> How many rook pieces can be put on a board so that no one can capture another? *Ans.* 8

> How many bishop pieces can be put on a board so that no one can capture another? *Ans.* 14

Answer the same question using knight pieces and using queen pieces.

Poker

It was the study of games of chance that first led Pascal and Fermat to their invention of probability. So it is not surprising that probability plays a role in a familiar card game, poker. Interesting classroom discussion can come from counting the different possible poker hands and computing various related probabilities, such as the following.

How many different 5-card poker hands can be dealt from a deck of 52 cards? The answer is the combination of 52 things taken 5 at a time:

$$_{52}C_5 = \binom{52}{5} = \frac{52!}{5!47!} = 2,598,960$$

This may not seem like very many. However, if you were able to deal out a different 5-card poker hand every second, working day and night, it would take about one month to deal them all.

Why is a flush better than a straight and a straight better than three of a kind in poker? The ranking of the poker hands is based upon the probabilities of their occurring—the better the hand, the lower the probability of its being dealt. These are the probabilities of the various hands:

straight flush	0.0000154	(1 in	64,974 hands)
four of a kind	0.0002401	(1 in	4,165 hands)
full house	0.0014406	(1 in	694 hands)
flush	0.0019654	(1 in	509 hands)
straight	0.0039246	(1 in	256 hands)
three of a kind	0.0211285	(1 in	48 hands)
two pair	0.0475390	(1 in	21 hands)
one pair	0.4225690	(1 in	$2\frac{1}{2}$ hands)
no pair	0.5011774	(1 in	2 hands)

Here is a detailed solution for the probability of a fairly common poker hand, one pair (the probabilities of the other hands can be found in much the same way):

$$\text{Probability of one pair} = \frac{\binom{13}{1}\binom{4}{2}\binom{12}{3}4^3}{\binom{52}{5}} = \frac{13 \cdot 6 \cdot 220 \cdot 64}{2,598,960} = 0.4225690$$

Roulette

Although it is quite probable that very few students have played roulette, it is a game that most have heard of. Bring a small roulette wheel into class and use it to introduce the concept of *mathematical expectation.*

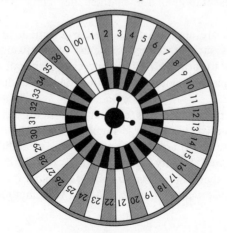

In games of chance, mathematical expectation can be thought of as the weighted mean of possible winnings, each weighted by its probability. Consider a $1 bet on a specific number on the roulette wheel.

The probability of winning is 1 out of 38. If you win, you get $35 plus your $1 back.

The probability of losing is 37 out of 38, and the loss is your $1.

$$E[X] = \frac{1}{38}(35) + \frac{37}{38}(-1) = \frac{35}{38} - \frac{37}{38} = -\frac{1}{19}$$

Your expected winnings are $-\frac{1}{19}$, or about a 5-cent loss on a $1 bet. Obviously, the game is weighted slightly in favor of the house.

If you're a conservative gambler, you might choose to bet on a color (red or black) instead of on a number. With 18 red and 18 black positions, surely your chances of winning are higher. If you win, you get only $1 plus your $1 back. However, one often forgets the two additional positions, 0 and 00, which are neither red nor black. Surprisingly, the expectation is exactly the same, a loss of about 5 cents per $1 bet.

$$E[X] = \frac{18}{38}(1) + \frac{20}{38}(-1) = \frac{18}{38} - \frac{20}{38} = -\frac{1}{19}$$

3.4   Enrichment topics

Many topics in mathematics are of interest in themselves, without regard for any particular application to the physical world or to other branches of mathematics. These are classified as enrichment topics and are most suitable for presentation in the mathematics class at a variety of grade levels.

These topics are appropriate to use as tangential to routine classroom activities and as such are effective means of motivation. Many of these can be expanded to form the basis of interesting laboratory lessons, whereas others lend themselves quite well to bulletin board displays and student projects.

As is true throughout this chapter, the items presented herein are merely representative of the large number of enrichment topics available for classroom use. Two especially good sources of additional ideas are the Twenty-seventh and Twenty-eighth Yearbooks of the National Council of Teachers of Mathematics, *Enrichment Mathematics for the Grades* and *Enrichment Mathematics for High School* (Washington, D.C.: N.C.T.M., 1963).

Polyominoes

A polyomino is merely a set of squares connected along their edges. The simplest form is a single square, called a *monomino*. Two connected squares are called a *domino,* three squares are called a *triomino,* and four connected squares are called a *tetromino.* In this enrichment topic we are concerned with the total number of possible arrangements of such figures that are not congruent to one another.

CLASSROOM PROCEDURES

Supply students with graph paper since this is the most convenient way of drawing and studying polyominoes. Together with the class, demonstrate the following figures and arrangements.

There is only one type of monomino and one domino.

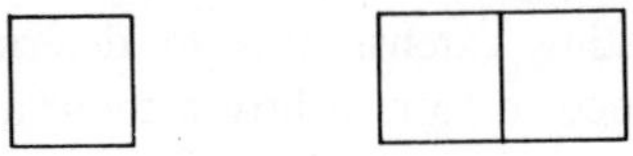

There are two possible arrangements using three squares.

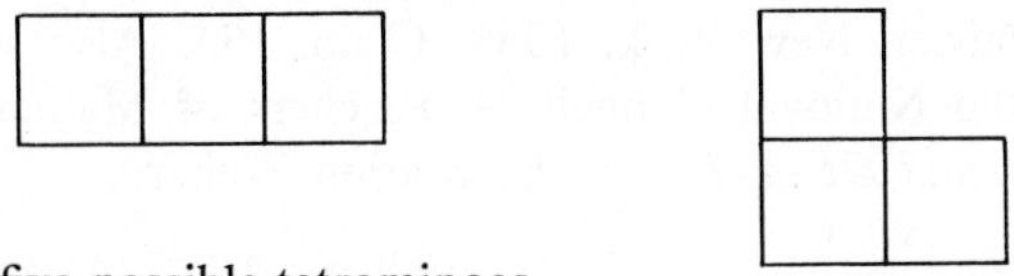

There are five possible tetrominoes.

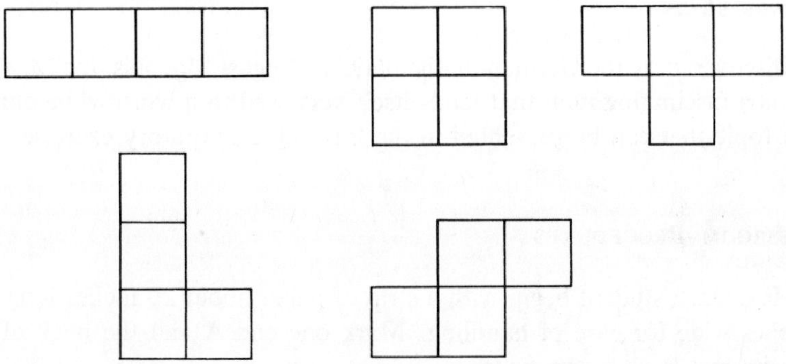

Having offered a class the preceding exposition, ask them to find all the possible pentominoes, figures formed by five connected squares. Caution them not to include any that are congruent to one another. For example, each of the following consists of just a single arrangement.

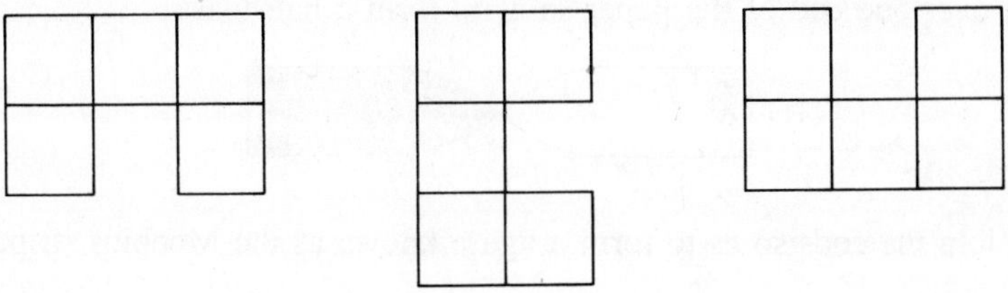

EXTENSIONS

An obvious extension is to have students search for all possible hexominoes, arrangements of six squares. However, this is a tedious and difficult project inasmuch as there are 35 different hexominoes. This could be a class project where each newly discovered hexomino is placed on the board until all 35 are found.

Another interesting, but difficult extension is to cut out all 12 possible pentominoes and try to arrange them to form a rectangle that is 5 units by 12 units in dimension. This is the basis for a recently marketed puzzle called *Hex*.

Yet another interesting extension is to determine which of the 12 pentominoes can be folded to form a box without a top.

For further reading on this topic see *Scientific American Book of Mathematical Puzzles and Diversions,* Martin Gardner, ed. (New York: Simon and Schuster, New York, 1959, Chap. 13). Also see the booklet published by the National Council of Teachers of Mathematics entitled *Boxes, Squares and Other Things,* by Marion Walter.

Moebius strips

Discovered by the German mathematician August Moebius, the Moebius strip is a fascinating item that lends itself very well to a worthwhile enrichment topic that can be presented in the form of a laboratory exercise.

CLASSROOM PROCEDURES

Have each student begin with a strip of paper about 20 inches long and 4 inches wide for ease of handling. Mark one end A and the back of the opposite end B as in the figure.

Turn over one end of the paper so as to form a half-twist.

Now join the ends so as to form a figure known as the Moebius strip.

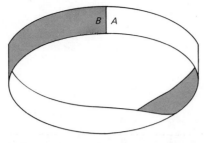

The Moebius strip is a one-sided figure. Start at *A* and draw a line down the middle of the strip. You will ultimately reach *B* without having to cross an edge, even though *B* was on the opposite side of the strip after making the initial half-twist. Now cut the figure down the middle. Instead of two figures as expected, you will end up with one band!

EXTENSIONS

Many extensions of this laboratory enrichment topic are possible. A few possibilities are listed.

1. Cut the newly formed figure down the middle again to see what type of figure is obtained, but first ask the class to predict the outcome.
2. Form a new Moebius strip, but this time cut along a line that is approximately one-third of the way across the band.
3. Form a band with two half-twists. Cut it down the center and discover the resulting figure.
4. Repeat the preceding extension, but begin with a band that has three half-twists.

Curve stitching

Curve stitching is an enrichment topic of particular interest because it seems to capture the attention of students of varying levels of ability. It is an effective topic to use before a holiday, and can serve as the basis for very dramatic bulletin board displays.

CLASSROOM PROCEDURES

Before allowing students to complete designs of their own, it is well to first have everyone construct one together under the teacher's guidance. A basic one with which to start begins by drawing an angle and marking off the same number of equally spaced units on each side of the angle. In the figure, 12 units are located and marked as shown on page 76.

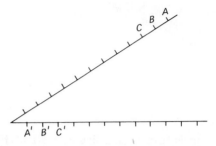

Next connect point A to point A', B to B', C to C', and so on. The line segments drawn will appear to form a curve called a *parabola*.

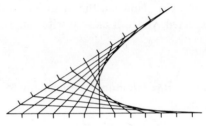

Many different variations are possible merely by changing the angle between the line segments, the distance between points, and by combining several curves. Here is an example of one possible variation.

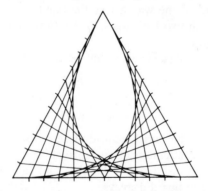

This enrichment topic is entitled "curve stitching" because the figures can be made by using colored thread and stitching through cardboard. Push up through each point from the back, and then stitch between points in the same manner as you would draw line segments.

EXTENSIONS

It is worthwhile to prepare a school exhibit consisting of a variety of curve-stitching designs made by individual students. A contest can be held with viewers asked to vote for the most attractive as well as the most original design created.

The figure shown is a far more difficult design to complete and may be suggested to some of the more industrious students in a class. The figure consists of a 24-sided polygon together with all its diagonals, giving the illusion of a series of concentric circles.

Tower of Hanoi

The Tower of Hanoi is a famous mathematical problem that students enjoy trying to solve. Although some students can easily construct display-size models for use in class, it can be played using three objects of different size, such as a quarter, nickel, and penny.

Classroom Procedures

Explain the rules for playing. The Tower of Hanoi puzzle consists of three disks, or pegs, of decreasing size, with the largest item on the bottom.

The object of the game is to transfer the disks to one of the other pegs, following these conditions:

1. Move only one disk at a time.
2. No disk may be placed on top of one smaller than itself.
3. Use the fewest possible moves.

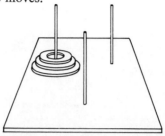

After explaining the rules of the game, students should be allowed to try to complete the game using three coins and three possible positions, as in the next figure. The three coins must be moved from position A to either B or C, using the rules given above. With three objects, only seven moves are needed.

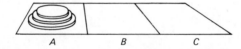

A   B   C

EXTENSIONS

After students are successful at completing the game in 7 moves, let them attempt the game using four objects. This can be tried using a quarter, nickel, penny, and dime. For four objects, and using the same rules, 15 steps are necessary. For five objects, 31 steps are needed. In general, for n objects, $2^n - 1$ steps are needed.

An ancient Hindu legend states that Brahma placed 64 disks of gold in the temple at Benares and called this the tower of Brahma. The priests were told to work continuously to transfer the disks from one pile to another in accordance with the rules set forth earlier. The legend states that the world would vanish when the last move was made. The minimum number of moves to complete this task is $2^{64} - 1$. Ask students to estimate how many moves this is and how long it would take at the rate of one move per second. It is interesting to note that $2^{64} - 1 = 18{,}446{,}744{,}073{,}709{,}551{,}615$; the world seems safe from destruction!

Tangrams

One of the oldest known puzzles is the ancient Chinese puzzle game of tangrams. Having amused and challenged people for thousands of years, it is certain to capture the interest of many students as well.

CLASSROOM PROCEDURES

Use a large square-shaped piece of cardboard, and complete the figure shown. To draw this figure, first locate points *M* and *N*, the midpoints of sides *AB* and *AD*. Draw *MN*. Then draw diagonal *BD* and part of the other diagonal of the square, shown as *PC*. Finally, draw *PQ* parallel to *AB* and *NR* parallel to *PC*.

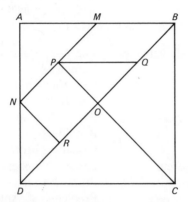

Next instruct the students to cut the figure along the lines drawn. This should give them seven separate pieces, consisting of five triangles, one square, and one parallelogram. A tangram puzzle now consists of arranging these seven pieces in the form of a given figure. Let the students try to use the seven pieces to form a triangle or a quadrilateral. Shown are two possible figures that can be formed from the seven pieces.

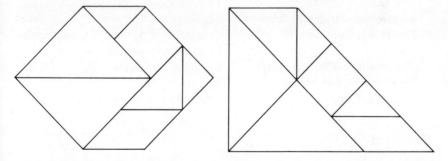

EXTENSIONS

A convex polygon is one with all its diagonals contained in the polygon and its interior. There are 13 possible tangrams altogether that are in the form of convex polygons, two of which are shown in the preceding figure. Of these 13 tangrams, one is a triangle, six are quadrilaterals, two are pentagons, and four are hexagons. Challenge your students to find all 13, and arrange these in a bulletin board display.

Suggested topics

There is an almost endless supply of topics that are suitable as enrichment topics for mathematics classes at various grade levels as well as at different levels of difficulty. Many of these can be assigned to individual students for research and for class projects and reports. Almost all of these are exceptionally well suited for mathematics club activities. Following is a list of some of the topics that are highly recommended for use in junior and senior high school mathematics classes.

Ancient Egyptian Mathematics	The Nine-Point Circle
Card Tricks	Nomographs
Conic Sections	The Number e
Diophantine Equations	The Number Pi
Divisibility of Numbers	Paper Folding
Euler's Formula	Paradoxes
The Euler Line	Pascal's Theorem
Fermat's Last Theorem	Perfect Numbers
Fibonacci Numbers	Prime Numbers
Finite Differences	Problems of Antiquity
Flatland	Projective Geometry
Flexagons	Pythagorean Triples
The Four-Color Problem	Regular Polygons
The Fourth Dimension	Regular Polyhedrons
Game Theory	Relativity
Geometry of Soap Bubbles	Shortcuts in Computation
The Golden Section	Tessellations
Koenigsberg Bridge Problem	Topology
Linear Programming	Trachtenberg System of Computation
Linkages	Transfinite Numbers
Mascheroni's Construction	Unsolved Problems
Mathematics and Music	Vectors
Napier's Rods	Zeno's Paradoxes

Exercises

1. Consider Example 7 on page 57. Using the digits 1 through 9, what is the largest sum that can be formed along *each* side of the triangle? What is the smallest sum possible?

2. For Exercise 1, try to construct a set of "magic" triangles so that the sum along each side is each of the numbers from the smallest to the largest possible number.

3. Consider a game whose objective is to reach the number 100. Two players alternate selecting one of the numbers 1 through 9. What is the strategy for winning such a game?

4. What is the appropriate move to guarantee a win for each of the problems of Tac Tix given on page 54.

5. There are 12 possible pentominoes. Find each one.

6. Determine which of the pentominoes found in Exercise 5 can be folded so as to form a box without a top.

7. There are 35 different hexominoes. Find as many of these as possible.

8. Find all 13 possible tangrams that form convex polygons, as suggested on page 79.

9. Many commercial games can be used in the mathematics class. One example is three-dimensional tic-tac-toe played on a 4 x 4 x 4 grid. How many different wins of 4 in a row are possible on such a grid?

```
            A
           ARA
          ARBRA
         ARBEBRA
        ARBEGEBRA
       ARBEGLGEBRA
      ARBEGLALGEBRA
       ARBEGLGEBRA
        ARBEGEBRA
         ARBEBRA
          ARBRA
           ARA
            A
```

10. Start with the A in the center and move to the right, or left, or up, or down, one letter at a time. In how many ways can you spell the word ALGEBRA?

Activities

1. Try to play each of the games described in Section 3.1 with a class of mathematics students. Report on your results in terms of student interest.

2. Search the literature and collect as many different mathematical games as you can find that are suitable for use in the classroom.

3. Prepare a collection of puzzles to supplement those given in Section 3.2.

4. Read the Twenty-seventh Yearbook of the National Council of Teachers of Mathematics, *Enrichment Mathematics for the Grades*. Prepare a report on three of the topics suggested in the book.

5. Repeat Exercise 4 for the Twenty-eight Yearbook, *Enrichment Mathematics for High School*.

6. Prepare a bulletin board display of curve-stitching designs.

7. Prepare a three-dimensional curve-stitching design.

8. Compile a list of as many different enrichment topics as you can think of, suggesting appropriate grade levels for each.

9. Prepare a one-period lesson plan using an enrichment topic appropriate for a junior high school mathematics class.

10. Repeat Activity 9 for a senior high school mathematics class.

References and selected readings

BAKST, AARON. *Mathematics, Its Magic and Mastery*. New York: Van Nostrand Reinhold Company, 1952.

BALL, W. W. R. *Mathematical Recreations and Essays*, 11th ed., rev. by H. S. M. Coxeter, 1939; reprinted New York: Macmillan Publishing Co., Inc., 1960.

BECK, ANATOLE, M. N. BLEICHER, and D. W. CROWE. *Excursions into Mathematics*. New York: Worth Publishers, Inc., 1969, Chap. 5, "Games."

GARDNER, MARTIN, ed. *Scientific American Book of Mathematical Puzzles and Diversions*. New York: Simon and Schuster, 1959.

———. *Second Scientific American Book of Mathematical Puzzles and Diversions*. New York: Simon and Schuster, 1961.

GOLOMB, SOLOMON. *Polyominoes*. New York: Charles Scribner's Sons, 1965.

HENDERSON, GEORGE L., and L. D. GLUNN. *Let's Play Games in Mathematics*, Vols. 1–8. Skokie, Ill.: National Textbook Company, 1971.

JOHNSON, DONOVAN A. *Games for Learning Mathematics*. Portland, Me.: J. Weston Walch, 1960.

KRAITCHIK, MAURICE. *Mathematical Recreations*. New York: Dover Publications, Inc., 1953.

McDONALD, JOHN DENNIS. *Strategy in Poker, Business, and War*. New York: W. W. Norton & Company, Inc., 1950.

MADACHY, JOSEPH S. *Mathematics on Vacation*. New York: Charles Scribner's Sons, 1966.

Mathematics Teacher. "Another Zip the Strip" (Nov. 1972), p. 669.

———. "A Card Trick" (May 1970), p. 395.

———. "The Cross Number Puzzle Solves a Teaching Problem" (Mar. 1969), p. 200.

———. "Instant Insanity" (Feb. 1972), p. 131.

———. "Mathematics on a Pool Table" (Mar. 1971), p. 255.

———. "Mathematics Word Search" (Apr. 1967), p. 357.

———. "Some Novel Möbius Strips" (Feb. 1972), p. 123.

———. "Three-Dimensional Tic-Tac-Toe" (Feb. 1971), p. 119.

———. "Zip the Strip" (Jan. 1971), p. 41.

National Council of Teachers of Mathematics. *Enrichment Mathematics for High School* (Twenty-eighth Yearbook). Washington, D.C.: N.C.T.M., 1963.

———. *Enrichment Mathematics for the Grades* (Twenty-seventh Yearbook). Washington, D.C.: N.C.T.M., 1963.

SCHAAF, WILLIAM L. *A Bibliography of Recreational Mathematics,* Vols. 1, 2. Washington, D.C.: National Council of Teachers of Mathematics, 1970.

Answers to section 3.2: puzzle pastimes

1. Start both hourglasses simultaneously. At the moment that the 7-minute hourglass is empty, the 11-minute one has 4 minutes left at the top of the glass. Start timing at this point; it will then take 4 minutes to complete. Then turn the 11-minute hourglass over and let it run its full 11 minutes; $4 + 11 = 15$.
2. Fill the 13-gallon can; pour out 5 gallons into the 5-gallon can, leaving 8 gallons. Pour this into the 11-gallon can. Repeat the procedure to obtain another 8 gallons in the 13-gallon can, leaving the last 8 gallons in the 24-gallon can.
3. Divide the coins into three sets of three each and weigh two sets. If they balance, the counterfeit coin is in the remaining set of three. Weigh two of these against each other; if they balance, the remaining coin is fake. If they do not balance, the fake coin is the one that shows to be lighter on the balance scale. In the original weighing, if the two sets of three coins do not balance, the fake coin is one of these that weigh less. In that case proceed as just described with the three coins.
4. Divide the coins into three sets of four each. In the first weighing balance a set of four against another set of four. This will determine which set of four coins contains the counterfeit one. In the second weighing balance two against two to determine which set of two coins contains the counterfeit. In the final weighing, balance one against one of these final coins to see which is lighter. Note that a much more difficult problem consists of determining the counterfeit coin in exactly

three weighings where you are not told whether the bad coin is lighter or heavier than the rest.

5. (a) 6   (b) 10   6. $15 and a pair of shoes.

7.

(4)

(8) (1)

(3)   (9)

(5) (2) (7) (6)

8.

| 7 | 2 | 9 |

| 8 | 6 | 4 |

| 3 | 10 | 5 |

9.

	3	5	
7	1	8	2
	4	6	

10–14. It is suggested that the reader attempt these problems by a trial-and-error process, using physical objects to experiment.

15. In the first step move 7 items from pile C to pile B. In the second step move 6 items from pile B to pile A. In step three move 4 items from pile A to pile C.

16. Many solutions are possible; the reader is encouraged to prepare these by means of actual experimentation.

17.

(a)   (b)   (c)   (d)

18. Slide the horizontal matchstick one-half its length to the right. Then move the upper left stick to the lower right position shown.

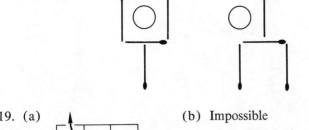

19. (a) (b) Impossible

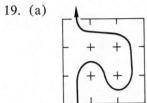

20. Yes 22.

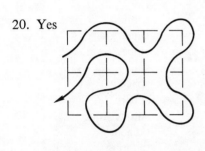

21. The first path can be traced in several ways; the second is not possible.

23. Pass a loop formed at the center of your string between the wrist and loop of the opponent and up over the hand itself. Then pull and the strings will be separated.

24. Make the center strip narrower than the punched hole. Then bend the strip through the hole, without tearing, and pull the string through.

25. Start in the upper right-hand corner. Move down, to the left, up, and to the right again in a spiral fashion until all symbols are covered. The repeating pattern in the array is triangle, circle, square, square, circle, circle.

26. (a) FRACTION   (b) MULTIPLY   (c) DECIMAL
    (d) SUBTRACT   (e) NUMBER   (f) DIVISION

27. (a) PASCAL   (b) DESCARTES   (c) EUCLID
    (d) PYTHAGORAS   (e) NEWTON   (f) ARCHIMEDES

28.   621   3862
       846   3450
      9071   412   (not unique)
     ------   ----
     10538

29. $108.24: $97.62
                 10.62

                $108.24

30. Some of the words included are:

acute	geometry	polygon
altitude	graphs	prism
arc	hexagon	proof
area	induction	radii
axiom	inscribe	radius
axis	intercept	ray
base	legs	similar
chord	length	sine
circle	limit	slope
complement	line	solid
cone	locus	space
conic sections	logic	sphere
deduction	mean	square root
definition	median	tangent
degree	midpoint	theorem
edge	origin	trapezoid
equal	parallel	triangle
equilateral	perpendicular bisector	volume
Euclid	pi	
extreme	plane	
face	point	

CHAPTER FOUR

LABORATORY EXPERIMENTS

A student needs to see and do things in the mathematics class. True there are definitions, skills, properties, postulates, and theorems to be learned. But there must also be exciting and challenging experiences where he can let his imagination and intuition roam, where he can manipulate objects as well as ideas, and where he can create and discover.

Renewed attention is now being given to the role of individual laboratory experiments and classroom laboratory activities as an integral part of the learning experience in mathematics. This chapter illustrates some of the many activities suitable for this approach. They are adaptable to a wide range of levels and abilities. Some are in the form of detailed worksheets or activity cards for the individual student to follow at his own pace. Some have specific objectives, directions, and analyses given and can be used as classroom demonstrations; others can be used as assignments or for special projects. At their best, laboratory experiments and activities should be varied in format, whether a familiar part of an individualized learning program or an occasional digression from routine teaching.

4.1 Magic-square activities

Many interesting activities involving exploration and arithmetic review can be presented within the format of magic squares. Each of the examples that follow is given in the form of a worksheet for individual use. However, these suggestions can be modified readily and extended to fit other needs.

Additional information of the topic of magic squares can be found in these references:

ANDREWS, W. S. *Magic Squares and Cubes*. New York: Dover Publications, Inc., 1960.

GARDNER, MARTIN, ed. *Second Scientific American Book of Puzzles and Diversions.* New York: Simon and Schuster, 1961.

KRAITCHIK, MAURICE. *Mathematical Recreations*. New York: Dover Publications, Inc., 1953.

Student worksheet 1:
Properties of magic squares

Magic squares are square arrays of numbers. They date back to ancient times when people believed they held strange mystic powers because of their special properties.

Add the numbers in each row, column, and diagonal in this magic square. What property do you discover?

Rows		
~~103~~	~~19~~	~~79~~
~~43~~	~~67~~	~~91~~
~~55~~	~~115~~	~~31~~

Columns		
103	19	79
43	67	91
55	115	31

Diagonals		
103	19	79
43	67	91
55	115	31

A square array of numbers is a magic square if the sums of the numbers in each row, column, and diagonal are the same. Find these sums for each of these magic squares.

7	2	16	9
12	13	3	6
1	8	10	15
14	11	5	4

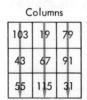

$2/3$	$1/12$	$1/2$
$1/4$	$5/12$	$7/12$
$1/3$	$3/4$	$1/6$

9	2	25	18	11
3	21	19	12	10
22	20	13	6	4
16	14	7	5	23
15	8	1	24	17

Study these square arrays. Then complete them so that they become magic squares.

16	2	12
	18	

8		14	3
15	2	9	4
	11		13

$1\frac{1}{6}$		
	$11\frac{1}{12}$	
1		$\frac{2}{3}$

Student worksheet 2:
Constructing odd-order magic squares

Here is a method that can be used to construct odd-ordered magic squares, those with an odd number of rows and columns.

Step 1: Write the first number, 1, in the middle of the top row. Fill in successive counting numbers moving diagonally upward and to the right. When the next position is off the square, enter the number at the opposite end of the next row or column. Note how this has been done for the entries 2 and 4.

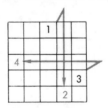

Step 2: Continue working diagonally upward and to the right. When you come to a position already occupied, enter the number in the space immediately below the last one filled. Note how this has been followed for the numbers 6 and 11.

Step 3: The space that follows the upper right-hand corner is the lower left-hand corner. Since this space is occupied by 11, the 16 is placed below 15 following the rule from step 2.

		1	8	15
	5	7	14	16
4	6	13		
10	12			3
11			2	9

Step 4: Continue the process until the 5-by-5 magic square is finished. A 5-by-5 magic square has 5 × 5, or 25, entries. Since 1 was the first number, 25 is the last.

17	24	1	8	15
23	5	7	14	16
4	6	13	20	22
10	12	19	21	3
11	18	25	2	9

Step 5: Check to see that the completed array is indeed a magic square by adding the numbers in each row, column, and diagonal.

Use the same method to complete these magic squares.

Start this magic square with a 1.

Start this 5 × 5 magic square with an 8.

Start this 7 × 7 magic square with a 1.

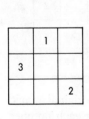

Check each array to be sure that it is a magic square. Then see if you can construct your own 9-by-9 magic square.

Student worksheet 3:
Rearranging entries in a magic square

Many different numbers can be used in forming a 3-by-3 magic square. Likewise, many arrangements can be made from the same set of numbers in a magic square.

8	1	6
3	5	7
4	9	2

This simple 3-by-3 magic square uses the numbers 1 through 9. The numbers in each row, column, and diagonal add to 15.

See if you can complete these magic squares using the same set of numbers, 1 through 9. Remember, the numbers in each row, column, and diagonal still must add to 15.

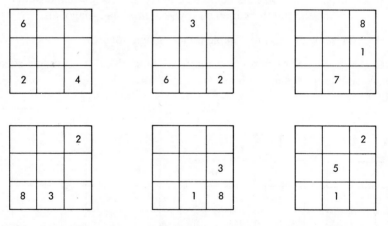

Cut out a square piece of paper and mark it with the original magic square. Turn the paper over and mark it on the back so that each number is in exactly the same place as in the front. Now see if you can get the magic squares above by simply rotating and flipping over this cutout. For example, if you flip over the original magic square about its vertical axis, you get the square in the upper left-hand corner. Find the one remaining arrangement possible with the numbers 1 through 9 that still forms a magic square.

Student worksheet 4:
Adding and multiplying with a constant

What happens when you add a constant to each entry in a magic square? Will you get another magic square?

5	6	1
0	4	8
7	2	3

Start with the magic square above. In each case add the constant given to each entry to form a new array. Then check to see if it, too, is a magic square.

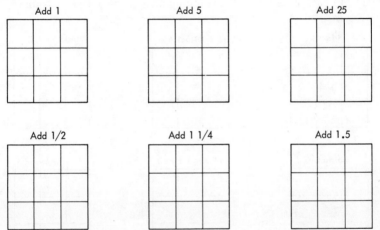

Add 1   Add 5   Add 25

Add 1/2   Add 1 1/4   Add 1.5

Does it appear that, when you add the same number to each entry in a magic square, another magic square is formed?

Now multiply each entry in the original magic square by these numbers.

Multiply by 9   Multiply by 3/4   Multiply by 1.3

Does it appear that, when you multiply each entry in a magic square by the same number, another magic square is formed?

Student worksheet 5:
Operating with integers

The entries in this magic square are integers. Show that the numbers in each row, column, and diagonal add to 0.

-3	2	1
4	0	-4
-1	-2	3

Start with the magic square given above and perform the operations indicated. In each case, see if the resulting array is also a magic square.

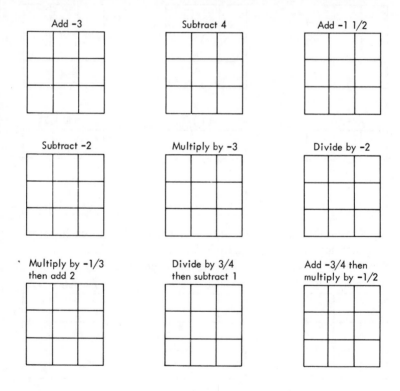

Add -3

Subtract 4

Add -1 1/2

Subtract -2

Multiply by -3

Divide by -2

Multiply by -1/3
then add 2

Divide by 3/4
then subtract 1

Add -3/4 then
multiply by -1/2

4.2 Geoboard activities

The geoboard can serve as an excellent aid for individual exploration and experimentation. It is so easy to use that students can become actively involved at a concrete level in the creative, imaginative aspects of geometric discovery. Classroom sets of geoboards are readily available commercially or they can be constructed by the school's shop classes with little difficulty. They usually consist of a board with a square array of nails about which rubber bands are stretched to form various shapes. Dot paper and grid paper are useful supplements for the geoboard and should be readily available to the student so that he can easily record specific figures when needed. A geoboard made of clear plastic can serve as a very effective and dynamic aid for the teacher when used with an overhead projector.

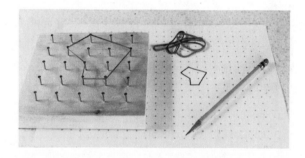

Many geometric concepts of the early elementary grades through the secondary years can be developed through geoboard activities and experiments. In this format they are often more exciting and motivating and assure a much higher degree of success. The suggestions given here are only representative of the wide and diversified variety of applications possible. Each student begins with a 5 x 5 geoboard, some rubber bands, a sheet of dot paper, and a set of activity cards to take him sequentially through the experiment. Answers may be placed on the back of the cards, if desired.

Geoboard laboratory experiment 1:
Shapes and sizes

STUDENT ACTIVITY CARDS

1. Form a square using the segment shown as a side.

2. Form another square, using this segment as a side.

3. Form yet another square with this segment as a side.

4. How many squares of different sizes can you form on the geoboard? Copy each one on dot paper.

5. How many isosceles triangles can you form using this segment as the base? Copy each one on dot paper.

6. How many isosceles triangles can you form using this segment as the base? Copy each one on dot paper.

7. How many rhombuses of different sizes can you form on the geoboard? Do not count those that are also squares. Copy each one you find on dot paper.

Geoboard laboratory experiment 2:
Congruent figures

STUDENT ACTIVITY CARDS

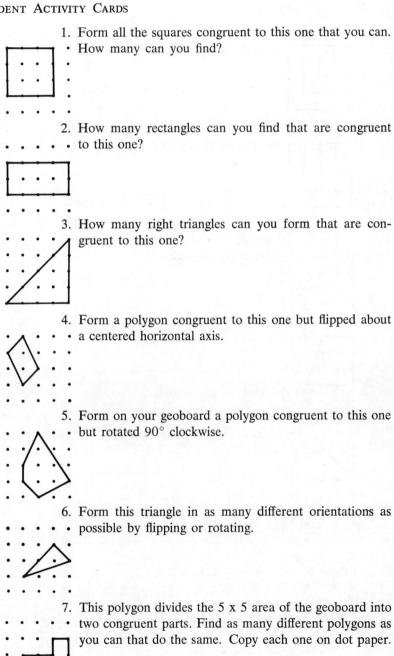

1. Form all the squares congruent to this one that you can. How many can you find?

2. How many rectangles can you find that are congruent to this one?

3. How many right triangles can you form that are congruent to this one?

4. Form a polygon congruent to this one but flipped about a centered horizontal axis.

5. Form on your geoboard a polygon congruent to this one but rotated 90° clockwise.

6. Form this triangle in as many different orientations as possible by flipping or rotating.

7. This polygon divides the 5 x 5 area of the geoboard into two congruent parts. Find as many different polygons as you can that do the same. Copy each one on dot paper.

Geoboard laboratory experiment 3:
Areas of polygons

STUDENT ACTIVITY CARDS

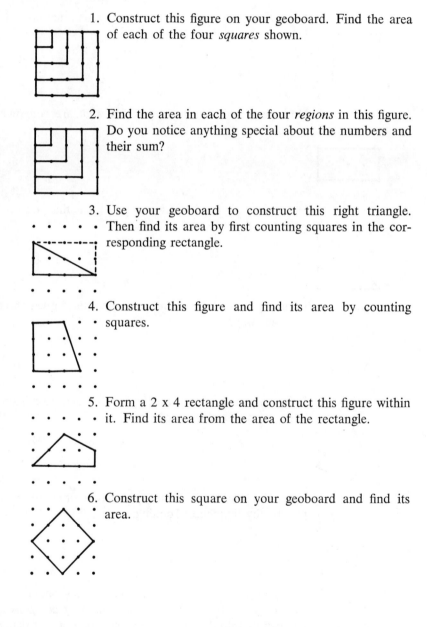

1. Construct this figure on your geoboard. Find the area of each of the four *squares* shown.

2. Find the area in each of the four *regions* in this figure. Do you notice anything special about the numbers and their sum?

3. Use your geoboard to construct this right triangle. Then find its area by first counting squares in the corresponding rectangle.

4. Construct this figure and find its area by counting squares.

5. Form a 2 x 4 rectangle and construct this figure within it. Find its area from the area of the rectangle.

6. Construct this square on your geoboard and find its area.

7. Construct this triangle and find its area.

8. Find the area of these triangles using your geoboard. What do you discover?

9. Construct this triangle on your geoboard and find its area. *Hint:* Start with the area of the rectangle enclosing it.

10. The area of the triangle on card 9 was $4\frac{1}{2}$. Construct as many other triangles as you can with the same area but with different shapes. Copy each one on dot paper.

11. Copy this polygon on your geoboard and find its area just by counting squares.

12. Make an intricate polygon on your geoboard and find its area. Then give it to a classmate and see how long it takes him to find the area.

Geoboard laboratory experiment 4:
Pick's formula

1. Copy this table on a piece of paper. Construct each polygon described on the remaining cards using the geoboard. Then enter the required value for b and i for that polygon in the table.

Card	b	i	Area

Let b be the number of points of the geoboard on the boundary of the polygon.

Let i be the number of points of the geoboard in its interior.

2. A 1 x 1 square

3. A 1 x 3 rectangle

4. A right triangle with legs 1 and 4

5. A right triangle with legs 1 and 3

6. Study the entries in the table thus far. Can you discover how to get A from b and i?

7. A 2 x 2 square

8. A 3 x 4 rectangle

9. A right triangle with legs 2 and 3

10. An isosceles triangle with two equal sides of 4

11. Study the entries in the table now. Can you discover how to get A from b and i?

12. Answer these questions from the table.

> Is b always greater than A?
> Is $b/2$ always greater than A?
> Is $b/2 + i$ always greater than A?
> How is A related to b and i?

13. Test your final answer for card 11 using this figure.

14. The formula relating the area A of a polygon to the number of points on the boundary b and in the interior i is called *Pick's formula:*

$$A = b/2 + i - 1$$

Were you able to discover it yourself?

4.3 Experiments with geometric models

Few topics are better suited for laboratory experiments than geometric models. The literature abounds with ideas and activities, some for the teacher, some for the class, and others for individual students. They seldom require more than a few sheets of heavy paper, a ruler, a pair of scissors, and some tape or glue. Ideally, these can be readily found in your mathematics laboratory in sufficient quantities for individual or group use.

The experiments described first are most appropriate for investigation by each individual student. But the teacher needs to become actively involved with students as they work through these experiments, offering helpful hints and suggestions where needed. It is also wise to have extra questions ready to challenge the better student to further exploration should he finish early.

Student worksheet 1:
Can you spot the cubes?

Many different patterns can be used to form models of cubes. Each pattern must have six squares for faces arranged so that, when assembled, no faces overlap.

1. Study the patterns given here and circle those that you think can be used for a cube. Be careful!
2. Check your answers by cutting out the patterns you are not sure of and try to assemble them.
3. Copy those that work on to a sheet of graph paper. Then draw as many more as you can find. Remember, count only those that are different patterns, not those that are different positions of the same pattern.

Student worksheet 2:
Which face is where?

All the patterns shown can be used to form a model of a cube. Assume that each cube has been assembled and properly positioned with the words on the outside and with the bottom face down and the back face in the rear.

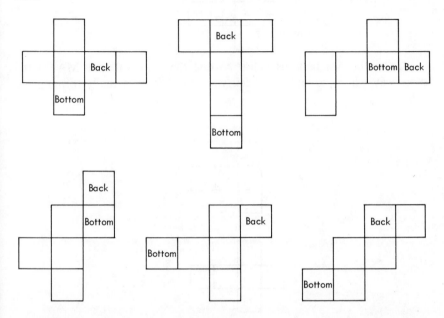

1. Mark the location of the remaining four faces on each pattern using this code.
2. Check each answer by cutting out and assembling each pattern.
3. Copy this pattern on a sheet of graph paper several times. Then see if you can find the four different ways you can label all the remaining faces so that the cube can still be properly assembled and positioned.

    F: front
    L: left
    R: right
    T: top

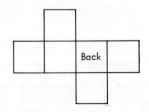

Student worksheet 3:
Can you spell MATH?

If the pattern below were assembled to form a cube, it would spell MATH around four of its faces.

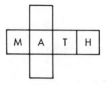

Letter the two patterns below so that they too will spell MATH the same way when assembled. Check your answers by actually forming the cubes.

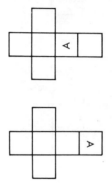

Letter these patterns so that each spells MATH when assembled. Cut out the patterns to check your answers.

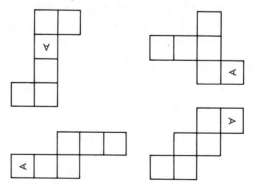

Student worksheet 4:
What do the pieces look like?

Imagine that this 3 x 3 x 3 cube is painted red and then cut into 27 small 1-inch cubes.

1. Complete the table describing how the 27 small cubes are painted.

number of faces painted	number of small cubes
0	
1	
2	
3	
4	
total	27

2. Suppose that a 4 x 4 x 4 cube were cut into 1-inch cubes the same way. How would the 64 small cubes be painted?

number of faces painted	number of small cubes
0	
1	
2	
3	
4	
total	64

3. If you had no trouble with these, try the same with an n x n x n cube. Be sure that your answers add to n^3.

Student worksheet 5:
What is magic about this cube?

Here is a pattern for a model of a 3 x 3 x 3 cube. When assembled, it appears to be formed from 27 smaller cubes, each bearing a number from 1 through 27

			10	24	8					
			23	7	12					
			9	11	22					
10	23	9	9	11	22	22	12	8	8	24	10
26	3	13	13	27	2	2	25	15	15	1	26
6	16	20	20	4	18	18	5	19	19	17	6
			20	4	18					
			16	21	5					
			6	17	19					

1. Cut out and assemble the cube.

2. Find the face with 1 on it. Write the three numbers in each of the two rows that contain the 1. Add each set of numbers. Are both sums the same? Now add the three numbers in each of the four other rows on the same face. Do you always get the same sum?

3. Find the face with 27 on it. Add each of the six rows of three numbers on this face. What do you find?

4. There are 36 different rows of three numbers on the six faces of the cube. Can you find them all? Are all their sums the same?

5. One of the numbers from 1 through 27 is not marked on any face. What number is it? Where is the cube located with this number on it?

6. The missing cube is numbered 14 and it is in the very middle of the 3 x 3 x 3 cube. There are 13 different rows and diagonals in the cube that contain this small middle cube numbered 14. Can you find them all? Are all their sums the same?

7. Why do you think this 3 x 3 x 3 cube is called a magic cube?

4.4   Euler's formula

Leonard Euler, Switzerland's most famous mathematician, lived from 1707 to 1783. (His name is pronounced like the word "oiler.") Among Euler's many contributions were the beginnings of topology, with his careful analysis of the famous Koenigsberg bridge problem, and his intriguing study of polyhedrons. Although known to Descartes more than 100 years earlier, this simple relationship among the number of vertices, faces, and edges of a polyhedron was independently discovered by Euler and now bears his name.

EULER'S   If V is the number of vertices, E the number of
FORMULA   edges, and F the number of faces of a simple
                         polyhedron, then $V + F = E + 2$.

The experiment described on this worksheet is designed for the individual student to follow in exploring these numbers, V, F, and E, for various polyhedrons. Hopefully, it will lead him to discover and support Euler's formula.

Student experiment 1:
Counting vertices, faces, and edges of polyhedrons

A polyhedron is a figure in space that has flat faces bounded by polygons. The prisms and pyramids shown here are examples of polyhedrons.

1. Count the number of vertices, faces, and edges for each prism and pyramid shown. Then record the numbers in the table.

Polyhedron	Number of		
	Vertices	Faces	Edges
cube tetrahedron pentagonal prism square pyramid hexagonal prism			

2. Study the values in the table. For each of these polyhedrons is the number of edges E greater than the number of vertices V and greater than

the number of faces F? Is E always less than the sum of V and F? Can you discover a relationship among the number of edges E and the number of vertices and faces V and F? Try to express this relationship in symbols.

3. Support your formula from step 2 using the V, F, and E from each of these polyhedrons.

Better students can investigate this topic further by studying one or more of these additional experiments.

Student experiment 2

Make a model of a cube. Now imagine a small piece cut off of each of the eight corners. Count the number of vertices, faces, and edges for the polyhedron thus formed. Do V, F, and E support Euler's formula?

Student experiment 3

Consider a prism with an n-gon (a polygon with n sides) for its bases. Express V, F, and E in terms of n and see if your results support Euler's formula.

Student experiment 4

Consider a pyramid with an n-gon as its base. Express V, F, and E in terms of n and see if your results support Euler's formula.

Student experiment 5

Patterns for models of the five regular polyhedrons—the tetrahedron, cube, octahedron, dodecahedron, and icosahedron—are given in Chapter 5. Construct these models; count the number of vertices, faces, and edges for each; and see if they, too, satisfy Euler's formula.

The next experiment leads to a discovery of Euler's relationship as it applies to figures on a plane; that is, the number of vertices plus the number of arcs is always two more than the number of regions:

$$V + A = R + 2$$

Student experiment 6:
Networks

A network on a plane consists of points called *vertices,* paths connecting them called *arcs,* and *regions* bounded by them. Some examples are shown here.

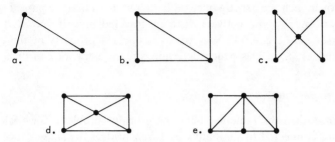

1. Count the number of vertices, regions, and arcs for each of the networks shown. Then record the numbers in the table.

Network	Number of		
	Vertices	Regions	Arcs
(a)			
(b)			
(c)			
(d)			
(e)			

2. For each of these networks is the number of arcs A ever less than the number of vertices V or the number of regions R? Is A less than the sum of V and R? Can you discover a relationship among the number of arcs A and the number of vertices and regions V and R? Try to express this relationship in symbols.
3. Now count V, R, and A for each of these networks and see if they support your formula from step 2.

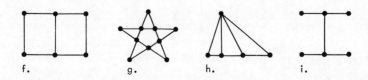

Better students can expand on this topic by considering some of the following experiments.

Student experiment 7

An odd vertex has an odd number of arcs coming from it; an even vertex has an even number. Study the number of odd and even vertices in each of the networks from step 1 of Experiment 6. Then see which networks can be traveled. To be able to travel a network, you must be able to cross over each arc exactly once, although vertices may be passed through more than once. Some networks can be traveled only if you start at the right vertex, so be careful. Now try to discover a property of the number of odd and even vertices which determines if the network can be traveled.

Student experiment 8

Euler studied the famous Koenigsberg bridge problem. Two islands in a river are connected to the banks by seven bridges, as shown. See if you can cross each bridge just once in taking a Sunday stroll.

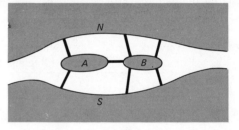

Student experiment 9

Euler's "solution" to the Koenigsberg bridge problem in Experiment 8 was that it could not be done because the corresponding network could not be traveled. Do you agree? Show how a bridge can be added so that they could all be traveled. Also show how one bridge can be removed so that the network can be traveled.

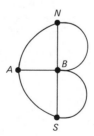

Student experiment 10

Investigate the properties of networks in space by studying the division of surfaces of solids.

4.5 Experiments in discovering number patterns

One of the objectives in teaching mathematics is to develop number awareness. Experiments that offer physical situations leading to the recognition of number patterns and sequences meet this need. They involve the individual student with a hands-on activity, a direct instruction leading to a countable result, a chance to discover a pattern and offer educated guesses at successive results before actual verification, and a challenging opportunity to generalize algebraically where appropriate. These are valuable ingredients in a meaningful and motivating laboratory experiment.

Consider the following experiment, which requires just some paper. Begin with a standard-sized sheet of paper. With zero folds it is 1 sheet thick. With one fold it is 2 sheets thick, and with two folds, 4 sheets thick. After showing these, have the student guess the thickness for three and four folds. Verify the answers by actually folding the paper. Then ask who has discovered the relationship between the number of folds and the thickness.

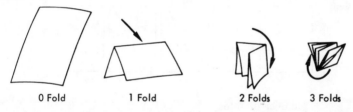

| 0 Fold | 1 Fold | 2 Folds | 3 Folds |

Have the students compute the number of folds needed to get the number of sheets in a 500-page book. (Note that a 500-page book uses only 250 sheets.) Give the paper to a student and see if he can fold it to this thickness. He will probably get no more than seven folds. Invariably, some student will suggest using a larger sheet of paper to get more folds. Be ready with a large page from a newspaper. Let the student try. The results will undoubtedly surprise both him and the rest of the class.

The examples of experiments in number patterns that follow include some with detailed descriptions and analyses and brief suggestions for others. They are probably best used one or two at a time throughout the year rather than all at one time. Some are more suited for lower grades and slower classes; others can be challenging for the better secondary students. But in all cases, the student should take educated guesses and try to verify them with the material supplied. Students should be encouraged to discover and verbalize the pattern. However, many of the generalizations of the nth terms can be left for the better students.

EXPERIMENT 1: Paper folding a number sequence

MATERIAL

One sheet of paper.

DIRECTIONS

1. Fold the paper once, open it up, and record the number of regions.
2. Fold again for the maximum number of regions possible.
3. Repeat the process again for three folds. Remember, open the paper flat before each new fold and always fold for the *maximum* number of regions possible.
4. Try to discover the number sequence and predict the result for four folds. Check your answer by folding again and counting regions.
5. Can you generalize the sequence for *n* folds?

ANALYSIS

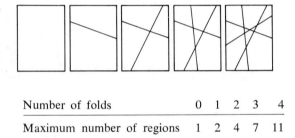

Number of folds	0	1	2	3	4
Maximum number of regions	1	2	4	7	11

NOTES TO THE TEACHER

1. The maximum number of regions is always formed if the new fold cuts every existing fold without passing through any intersection point.
2. Generalization for *n* folds:

The first differences form the counting numbers. Hence the *n*th fold produces the sum of the first *n* counting number plus 1:

1 ‿ 2 ‿ 4 ‿ 7 ‿ 11 ‿ 16
  1 2 3 4 5

$$\frac{n(n+1)}{2} + 1$$

EXPERIMENT 2: Number patterns from cutting string

MATERIAL

Several pieces of string about 2 feet long and a pair of scissors.

DIRECTIONS

1. If you know the number of cuts in a string, do you know the number of pieces? Try completing the table without cutting the string first.

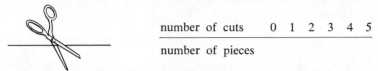

number of cuts	0	1	2	3	4	5
number of pieces						

If there are *n* cuts in the string, how many pieces will there be?

2. Fold the string as shown before you cut. Do you know the number of pieces that will be formed this way from a given number of cuts? Again, try to complete the table without cutting the string.

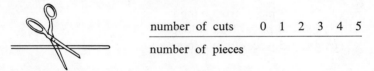

number of cuts	0	1	2	3	4	5
number of pieces						

If there are *n* cuts, how many pieces will there be?

3. Loop the string through the scissors as shown. Find how many pieces you get for each number of loops.

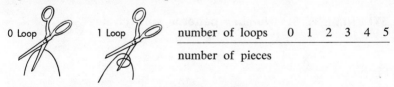

number of loops	0	1	2	3	4	5
number of pieces						

If there are *n* loops, how many pieces will there be?

4. Tie the ends of the string together and then loop it through the scissors. How many pieces do you get for each number of loops? Cut some and try. How many pieces will you get from *n* loops?

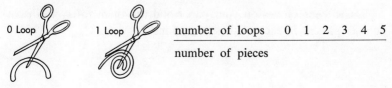

number of loops	0	1	2	3	4	5
number of pieces						

5. How many pieces will you get if the string is folded in each of these ways before being looped around the scissors? Guess your answers for 0, 1, 2, and 3 loops. Verify your answers by actually cutting, then predict the answer for *n* loops.

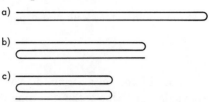

a)

b)

c)

ANALYSIS

In each case, the actual number of places the string is cut is the clue to the answer.

1. *n* + 1 pieces from *n* cuts

2. 2*n* + 1 pieces from *n* cuts

3. *n* + 2 pieces from *n* loops

4. 2*n* + 2 pieces from *n* loops

5. (a) 2*n* + 3 pieces from *n* loops
   (b) 3*n* + 4 pieces from *n* loops
   (c) 4*n* + 5 pieces from *n* loops

EXPERIMENT 3:   Number patterns with circles

MATERIAL

Compass, ruler, and paper.

DIRECTIONS

Draw some circles on your paper and use them to try to discover the number pattern in each of these situations.

1. Find the maximum number of regions possible in a circle for various numbers of radii, diameters, and chords.

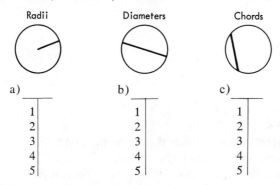

Radii	Diameters	Chords
a)	b)	c)
1	1	1
2	2	2
3	3	3
4	4	4
5	5	5

Find the maximum number of regions possible in a plane for various numbers of concentric circles, tangents, and secants.

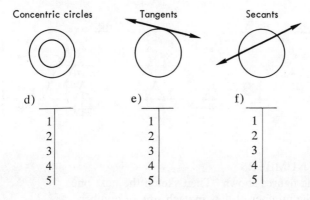

Concentric circles	Tangents	Secants
d)	e)	f)
1	1	1
2	2	2
3	3	3
4	4	4
5	5	5

2. Study each pattern and see if you can find how to predict the next term.
3. Find the corresponding number for each problem given n items.

ANALYSIS

a) n regions for n radii
b) $2n$ regions for n diameters
c) $n(n + 1)/2 + 1$ regions for n chords
d) $n + 1$ regions for n circles
e) $(n + 1)(n + 2)/2$ regions for n tangents
f) $(n + 2)(n + 3)/2 - 2$ regions for n secants

EXPERIMENT 4: Number patterns in figurate numbers

MATERIAL

Graph paper.

DIRECTIONS

TRIANGULAR NUMBERS
1. Copy the figures shown. Then sketch the next one.
2. Count the dots in each triangular number.
3. Complete the first differences. What pattern do you get?
4. Complete the second differences. What do you discover?
5. What is the nth triangular number?

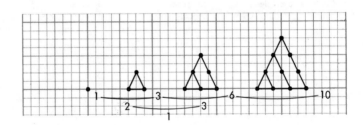

SQUARE NUMBERS
1. Copy the figures shown. Then sketch the next one.
2. Count the number of dots in each square number.
3. Complete the first and second differences. What do you discover?
4. What is the nth square number?

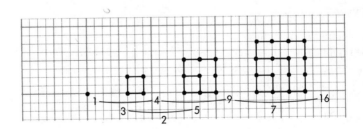

PENTAGONAL NUMBERS
1. Copy the figures shown. Then sketch the next pentagonal number.
2. Count the number of dots in each pentagonal number.
3. Complete the first and second differences. What do you discover?
4. What is the nth pentagonal number?

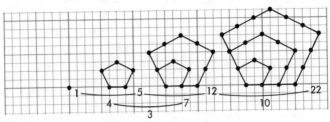

ANALYSIS

1. Point out the similar figures in the figurate number sketches.
2. Note that the second differences for these figurate numbers are constant.
3. nth triangular number: $n(n + 1)/2$
   nth square number: n^2
   nth pentagonal number: $3n^2/2 - n/2$

EXTENSIONS
Students might be interested in exploring similar properties of other figurate numbers.

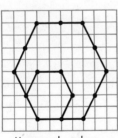

Hexagonal numbers

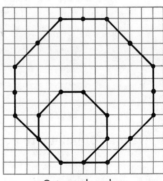

Octagonal numbers

EXPERIMENT 5: Counting squares

MATERIAL

Some square pieces of paper.

DIRECTIONS

Repeatedly divide a square into smaller and smaller squares. Count the number of the smallest squares formed after each successive division.

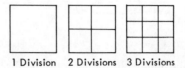

1 Division   2 Divisions   3 Divisions

Repeat the process only this time count all squares of all sizes.

ANSWERS

number of divisions	1	2	3	4	5
number of small squares	1	4	9	16	25

number of divisions	1	2	3	4	5
number of all squares	1	5	14	30	55

EXPERIMENT 6: Counting regions

MATERIAL

A piece of string.

DIRECTIONS

Place the string on the desk so that it crosses itself just once. How many regions are formed? Move the string so that it crosses itself twice.

118

Now how many regions are formed? Is there any other way that you can lay the string so it crosses itself twice but forms a different number of regions? Try the same for three and four crossings. Can you discover a pattern?

EXPERIMENT 7: Counting triangles

MATERIAL

A triangular piece of paper.

DIRECTIONS

Repeatedly fold a triangle through one of its vertices. Count the total number of triangles formed after each fold. Try to discover the pattern.

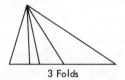

3 Folds

ANSWERS

number of folds	0	1	2	3	4	5
number of triangles	1	3	6	10	15	21

4.6 Algebraic identities through paper folding

Here are two experiments that illustrate algebraic properties as a result of paper-folding activities. Students often need visual, concrete examples to help reinforce abstract algebraic concepts. From a historical point of view, the ancient Greeks solved algebraic problems by geometric means such as these 1000 years before the invention of algebra as we know it today.

EXPERIMENT 1: $(a + b)^2 = a^2 + 2ab + b^2$

OBJECTIVE

To give a geometric demonstration for $(a + b)^2 = a^2 + 2ab + b^2$.

MATERIAL

One square piece of paper.

DIRECTIONS

1. Begin with a square sheet of paper.

2. Fold one edge over at a point E to form a vertical crease parallel to the edge. Label the longer and shorter dimensions a and b.

3. Fold the upper right-hand corner over onto the crease to locate point F. Folding this way, point F will be the same distance from the corner as point E.

4. Now fold a horizontal crease through F and label all outside dimensions.

5. Note that two square regions and two rectangular regions are formed.

area of square Y: a^2
area of square X: b^2
area of rectangle W: ab
area of rectangle Z: ab

6. Show that these areas together must equal $(a + b)^2$.

120

ANALYSIS

The original square paper measures $a + b$ on each edge and hence has an area of $(a + b)^2$. The four smaller parts combine to give an area of $a^2 + 2ab + b^2$. But together their areas must equal that of the original square. Therefore,

$$(a + b)^2 = a^2 + 2ab + b^2$$

EXPERIMENT 2: $(a + b)(a - b) = a^2 - b^2$

OBJECTIVE

To give a geometric demonstration for $(a + b)(a - b) = a^2 - b^2$.

MATERIAL

One rectangular sheet of paper.

DIRECTIONS

1. Begin with a rectangular sheet of paper.

2. Fold the left edge down onto the bottom edge to locate point P.

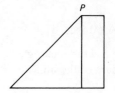

3. Fold vertically through point P. This forms a square on the left. Label the longer and shorter dimensions a and b.

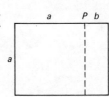

4. Fold the upper right-hand corner over on the crease to locate point Q.

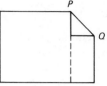

5. Now fold a horizontal crease through Q and label all outside dimensions.

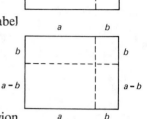

6. Note the three rectangular and one square region formed.

area of square X: b^2
area of rectangle W: ab
area of rectangle Y: $a(a - b) = a^2 - ab$
area of rectangle Z: $b(a - b) = ab - b^2$

7. Show that the areas of Y and Z together equal both $(a + b)(a - b)$ and $a^2 - b^2$.

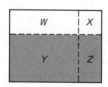

ANALYSIS 1

This rectangle has an area $(a + b)(a - b)$. But its area is also the sum of the areas of Y and Z.

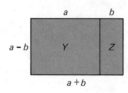

area $Y = a(a - b) = a^2 - ab$
area $Z = b(a - b) = ab - b^2$
sum of areas $\qquad \overline{a^2 - b^2}$

Hence $(a + b)(a - b) = a^2 - b^2$.

Analysis 2

Arrange pieces X, Y, and Z to form a square with area a^2.

Removing piece X gives an area of $a^2 - b^2$. But pieces Y and Z can be rearranged to form a rectangle with area $(a + b)(a - b)$.

Hence $a^2 - b^2 = (a + b)(a - b)$.

4.7   Geometric properties through paper folding

Many geometric properties can be effectively and dynamically illustrated through paper-folding activities. Several are described here. Others can be found in the following references:

Johnson, Donovan A. *Paper Folding for the Mathematics Class.* Washington, D.C.: National Council of Teachers of Mathematics, 1957.

Yates, Robert C. *Paper Folding* (Eighteenth Yearbook). Washington, D.C.: National Council of Teachers of Mathematics, 1945.

Students will enjoy folding their own examples and making their own discoveries. Vivid examples can be illustrated on the overhead projector by the teacher using creases in waxed paper.

EXPERIMENT 1: Angles of a triangle

O<small>BJECTIVE</small>

To show that the sum of the measures of the angles of a triangle is 180°.

M<small>ATERIAL</small>

A triangular piece of paper.

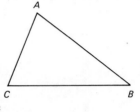

D<small>IRECTIONS</small>

1. Begin with a triangular region *ABC*.
2. Fold vertex *C* onto side *BC* so that the crease passes through vertex *A*.
3. The crease *AD* is the altitude through vertex *A*.
4. Fold vertex *A* onto point *D*.
   Fold vertex *C* onto point *D*.
   Fold vertex *B* onto point *D*.
5. Show that the sum of the measures of the angles *A*, *B*, and *C* is 180°.

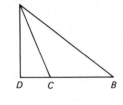

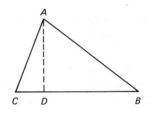

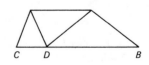

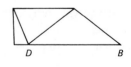

ANALYSIS

The three angles have been folded onto a straight-line segment. Hence the sum of their measures must be 180.

EXPERIMENT 2: Area of a triangle

OBJECTIVE

To show that the formula for the area of a triangle is

$$A = \tfrac{1}{2}bh$$

MATERIAL

Use the folded triangle from Experiment 1.

DIRECTIONS

1. Note that twice the area of the rectangle is the area of the triangle.

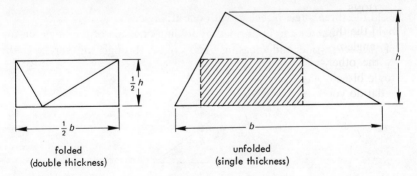

folded
(double thickness)

unfolded
(single thickness)

2. The length of the rectangle is half the base of the triangle. The width of the rectangle is half the height of the triangle. Show that the area of the triangle is $\tfrac{1}{2}bh$.

ANALYSIS

$$\text{area of triangle} = 2(\text{area of rectangle}) = 2(l \times w)$$
$$= 2(\tfrac{1}{2}b \times \tfrac{1}{2}h)$$
$$= 2(\tfrac{1}{4}bh)$$
$$= \tfrac{1}{2}bh$$

EXPERIMENT 3:   Angle bisectors, altitudes, and medians of triangles

OBJECTIVES

To show that the angle bisectors, the altitudes, and the medians of a triangle are concurrent.

MATERIAL

Three triangular pieces of paper.

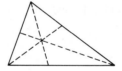

DIRECTIONS

1. Fold the three angle bisectors with one triangle.
   An angle bisector can be formed by folding one side of the angle on top of the other side and creasing. The crease through the vertex is the angle bisector.
   What do you discover?

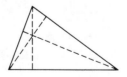

2. Fold the three altitudes using another triangle.
   An altitude from a vertex can be formed by folding the opposite side

upon itself such that the crease passes through the given vertex.
What do you discover?

3. With the third triangle fold the three medians.
   To fold a median to a given side, first bisect the side. This can be done
   by folding it upon itself such that the endpoints coincide. Then fold a
   crease through this midpoint and the opposite vertex.
   What do you discover?

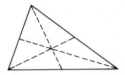

ANALYSIS

The angle bisectors of a triangle are concurrent as are the altitudes and
medians. The common point for the angle bisectors and medians will always lie within the triangle. However, the triangle must be acute for the
altitudes to meet within it.

4.8 Probability experiments

Experiments in probability can be exciting and interesting to students with
a wide range of ages and abilities. The element of chance and doubt, the
close association with familiar games, and the active involvement by the individual student all help to make the topic motivating. Here is an example
of a game that might be used in class to introduce the topic.

Each student draws a number line from —5 through 5. Then starting at 0 he moves one unit at a time to the right or left depending upon whether a tossed coin lands heads or tails. The first one to reach —5 or 5 tossing his own coin wins.

Reaching a 5 requires 5 heads; reaching —5 requires 5 tails. In 5 tosses the probability of all heads is $\frac{1}{32}$, as it is for all tails. So if you allow just 5 tosses, the chances are 1 in 16 that a student will reach —5 or 5. In an average class, then, at least one student should be expected to win in just 5 tosses.

An interesting follow-up is to find how many in class landed at each number and compare these results with those from a list of all 32 ways in which the 5 tosses could fall.

probabilities

1 way to stop at —5

TTTTT $\qquad \frac{1}{32} = .03125$

5 ways to stop at —3

TTTTH TTTHT TTHTT THTTT HTTTT $\qquad \frac{5}{32} = .15625$

10 ways to stop at —1

TTTHH TTHTH THTTH HTTTH TTHHT
THTHT HTTHT THHTT HTHTT HHTTT $\qquad \frac{10}{32} = .31250$

10 ways to stop at   1

HHHTT HHTHT HTHHT THHHT HHTTH
HTHTH THHTH HTTHH THTHH TTHHH $\qquad \frac{10}{32} = .31250$

5 ways to stop at   3

HHHHT HHHTH HHTHH HTHHH THHHH $\qquad \frac{5}{32} = .15625$

1 way to stop at   5

HHHHH $\qquad \frac{1}{32} = .03125$

$\overline{}$

1.00000

Many simple activities involve the prediction of probabilities, collection of experimental data, and the analysis of results.

What is the probability that a letter chosen at random from a newspaper is an E?

What is the probability that a randomly chosen word from a newspaper has less than five letters?

Sample the letters and words in a paragraph and see.

What is the probability that the last digit in a telephone number is 0 through 4?

What is the probability that the first digit after the exchange number is 0 through 4?

Sample 100 numbers in a telephone directory and see.

Other experiments estimate probabilities that can then be computed mathematically.

Coin tossing:	Heads on a single toss	1/2
	Two heads on two tosses	1/4
	Three alike on three tosses	1/4
Cutting a deck of cards:	Red	1/2
	Diamonds	1/4
	Face card (J, Q, K)	3/13
Rolling dice:	Even number on a single die	1/2
	Two alike with a pair of dice	1/6
	7 or 11 with a pair of dice	2/9

Experiment 1 describes the format that can be used in class for such an activity. Students gain experience in collecting and recording data, in maintaining computational skills with fractions and per cents, in the construction and interpretation of graphs, and in understanding the meaning of probability *in the long run*.

EXPERIMENT 1: Coin tossing

What is the probability that when two coins are tossed, both fall heads? Repeated tosses emphasize the notion of probability *in the long run*. As more trials are performed the computed experimental probabilities tend to level off at the correct value. This is vividly illustrated by the line graph and by the cumulative fractions and per cents in the table.

Suggested Procedures

Students first guess at the answers. Then tossing two coins at a time, they compute the ratio of successes (2 heads) to trials and the corresponding cumulative per cents. These per cents are then plotted on a graph for each of 20 successive trials. Typical results might look as here:

Toss	1	2	3	4	5	6	7	8	9	10	11	12	13	14	15	16	17	18	19	20
Two Heads	X					X							X	X						X
Not Two Heads			X	X	X	X		X	X	X	X	X		X		X	X	X	X	
Rate of Successes To Total	$\frac{1}{1}$	$\frac{1}{2}$	$\frac{1}{3}$	$\frac{1}{4}$	$\frac{1}{5}$	$\frac{1}{6}$	$\frac{2}{7}$	$\frac{2}{8}$	$\frac{2}{9}$	$\frac{2}{10}$	$\frac{2}{11}$	$\frac{2}{12}$	$\frac{3}{13}$	$\frac{3}{14}$	$\frac{4}{15}$	$\frac{4}{16}$	$\frac{4}{17}$	$\frac{4}{18}$	$\frac{4}{19}$	$\frac{5}{20}$
Per Cent of Successes	1.000	.500	.333	.250	.200	.100	.285	.250	.222	.200	.181	.166	.230	.214	.266	.250	.235	.222	.210	.250

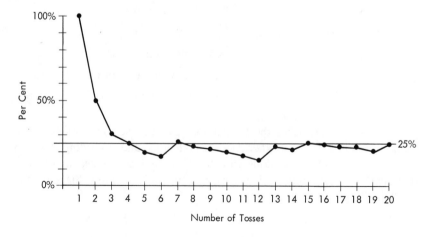

The original guesses can then be compared with the collected data and modified, based upon the experimental evidence.

ANALYSIS

Students may first guess the probability to be $\frac{1}{3}$, arguing that the coins can fall 2 heads, 1 head, or 0 heads. However, the correct answer is $\frac{1}{4}$, since only one of these four equally likely possible outcomes is a success.

$$HH \quad HT \quad TH \quad TT$$

The greater the number of repetitions, the more likely the graph will tend toward this value.

EXPERIMENT 2: Tossing tacks

What is the probability that a tack lands point up when dropped? In this experiment the probabilities of various tacks falling point up are estimated by experimentation.

SUGGESTED PROCEDURES

Three different types of tacks and thumbtacks are selected. Students first study their construction and try to guess which is most likely to fall point up and which least likely.

Ten tacks of one type are dropped on a flat surface 10 different times. Then the probability that a single tack would land point up is estimated using this ratio:

$$P(\text{up}) = \frac{\text{number falling point up}}{100}$$

The experiment is repeated with the other two tacks and then final results are compared with the original guesses.

ANALYSIS

Experimentation is the best way to estimate probabilities here since no equivalent simple mathematical model can be designed. Some of the characteristics that increase the probability of landing point up are larger heads, flatter heads, heavier heads, and shorter points.

EXPERIMENT 3: Tossing pennies on squares

What is the probability that a penny tossed on a $1\frac{1}{2}$-inch ruled grid will not fall on a line?

Repeated experimentation supplies initial data for an estimate, and careful mathematical analysis yields the correct value.

SUGGESTED PROCEDURES

Students first guess at what the probability will be. A large sheet of paper is then ruled with $1\frac{1}{2}$-inch squares and 10 pennies tossed on the grid 10 separate times. The number N of pennies that do not fall on a line is counted and the probability estimated using this ratio:

$$P = \frac{N}{100}$$

Answers are then compared with original guesses.

ANALYSIS

An accurate probability can be determined by comparing areas. Assume that the center of the penny lands anywhere within a given square. The radius of a penny is $\frac{3}{8}$ inch, so if its center lies more than $\frac{3}{8}$ inch from any side of the square, the penny will not be on a line. Hence the only possible location for the center of the penny such that the penny itself is not on a line would be within a small $\frac{3}{4}$-inch square. Compare this area to the total for the original square for the correct probability.

$$P(\text{not on line}) = \frac{\frac{3}{4} \times \frac{3}{4}}{1\frac{1}{2} \times 1\frac{1}{2}} = \frac{1}{4}$$

EXPERIMENT 4: Buffon's needle problem

Here is a famous problem which was first presented by Count Buffon in the 1700s. Needles are dropped in a special way on a ruled surface such that the computed probability that one falls on a line gives an estimate involving π.

SUGGESTED PROCEDURES

Toothpicks are cut to a length of 1 inch and a surface ruled with lines 2 inches apart. The toothpicks are randomly and repeatedly dropped from a reasonable height onto the ruled surface. The number falling on a line are counted along with the total number dropped. Their ratio should approximate $1/\pi$.

$$\text{To 10 places} \quad \pi = 3.1415926535 \ldots$$
$$1/\pi = .3183098861 \ldots$$

The latter value is always approximated as long as the distance between the lines is twice that of the length of the toothpicks dropped. Increasing the number of trials tends to improve the approximation.

ANALYSIS

A detailed analysis of this problem requires calculus but yields the *exact* result, $1/\pi$.

EXPERIMENT 5: The birthday problem

What are the chances that at least two people in a crowd have the same birthday?

Surveys give data for initial experimental estimates to a problem that has an exact mathematical analysis.

SUGGESTED PROCEDURES

The problem is first carefully discussed. Students should guess what they think are the chances of a duplication with groups of say 15, 25, and

133

60. Next various groups are surveyed and records kept of the number of duplicate birthdays found in each group. The class itself should be surveyed first. Then fellow students, teachers in the school, parents, or families in the neighborhood can be surveyed along with other recorded data such as the birthdays of all the presidents of the United States. Finally, modifications on the original estimates can be made based upon the collected data.

ANALYSIS

Omitting leap year's February 29, there are 365 possible birthdates. With just two people, the chances are 1 in 365 that they have the same birthday. However, with 3 people, the chances are greater since *any* two could have the same date. The greater the crowd, the greater the chance of a duplication of birthdates.

Here are the results for various group sizes.

Size of group	15	18	23	40	60
Chances of a duplication	better than 1 in 4	better than 1 in 3	better than 1 in 2	better than 9 in 10	almost certain
Approximate probability	1/4	1/3	1/2	9/10	near 1

For n people, the exact probability of at least one duplication can be found using the formula

$$1 - \frac{365}{365} \cdot \frac{364}{365} \cdot \frac{363}{365} \cdot \cdot \cdot \frac{365 - n + 1}{365}$$

Computing these probabilities can be an interesting exercise.

4.9   Constructing the conic sections

One of the most important sets of curves in mathematics is the conics. The ancient Greek Apollonius (225 B.C.) wrote a treatise entitled *Conic Sections,* in which he described how the ellipse, parabola, and hyperbola can be formed by passing a plane through a cone at various angles. Models illustrating these are familiar to most mathematics teachers. However, students also should be shown how a single cutting plane can be moved at different angles to form the various conics.

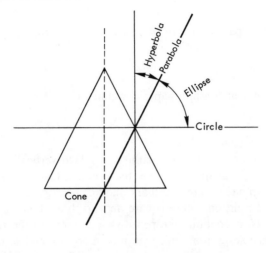

A loop of string and two tacks on a drawing board or two suction cups on the chalkboard are all that are needed to draw an ellipse. Changing the length of the loop or the distance between the tacks changes the shape of the ellipse.

Although most algebra courses include some treatment of the conics from an analytic approach, little is usually done with them at a more informal level. The following experiments illustrate some simple methods for constructing the conics. They offer interesting and valuable activities for students at both the junior and senior high levels.

EXPERIMENT 1:   Parabolas on graph paper

OBJECTIVE

To show how parabolas can be constructed on graph paper.

MATERIAL

One sheet of ¼-inch graph paper per student.

DIRECTIONS

1. Draw a line down the middle of the paper held vertically.
2. Mark a point P on this line near the bottom. Count to the right one square and up one square and mark another point.
3. Count to the right one more square and up three more squares for the next point. Now continue the process—one more to the right and five more up, then seven more up, then nine more up, and so on.
4. Repeat the same process starting with point P but this time moving to the left and up.
5. Finally, connect the points with a smooth curve, a parabola.

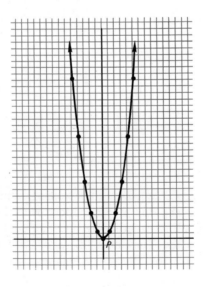

NOTES

1. If you place an x- and a y-axis through the point P, the equation of the parabola drawn is $y = x^2$.

x	0	1	2	3	4	5
y	0	1	4	9	16	25

However, in drawing the parabola following the directions above, the sums of the odd numbers are used to produce the squares.

$$1 = 1$$
$$4 = 1 + 3$$
$$9 = 1 + 3 + 5$$
$$16 = 1 + 3 + 5 + 7$$
$$25 = 1 + 3 + 5 + 7 + 9$$

2. A whole family of parabolas through P can be drawn. Just move some other number of units to the left and right each time, such as $\frac{1}{2}$, 2, or 3.

EXPERIMENT 2: Conics from circles and lines

OBJECTIVE

To construct the conics using the basic loci definitions.

MATERIAL

A set of worksheets with concentric circles and parallel lines as shown.

DIRECTIONS

PARABOLA: Worksheet 1

1. Locate and mark the intersection of circle 1 and line 1. Next mark the two intersections of circle 2 and line 2. Then repeat for circle 3 and

line 3, circle 4 and line 4, and so on. Connect these points with a smooth curve, a parabola.

Take any point on the curve. Is it the same distance from line 0 and the center of the circles?

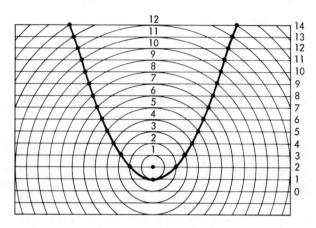

ELLIPSE: Worksheet 2

2. Locate and mark the points where circles numbered 4 and 8 intersect. Each of these four points is 4 units from one center and 8 units from the other. The sum of the distances from the two centers is 4 + 8, or 12. Next mark the four intersection points of circles 5 and 7. They, too, are located such that the sum of their distances from the two centers is 12. Now mark all other sets of points where the sum of the distances from the two centers is 12. For example, 3 and 9 and 2 and 10. Connect the points with a smooth curve, an ellipse.

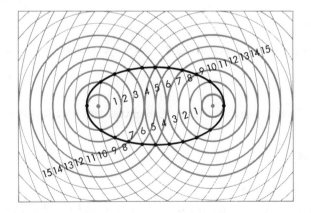

HYPERBOLA: Worksheet 2

3. Locate and mark points in a similar fashion, but this time with those that always have a difference of 6. For example, 8 and 2, 9 and 3, 10 and 4, and so on. The two smooth curves formed by connecting these sets of points are branches of a hyperbola.

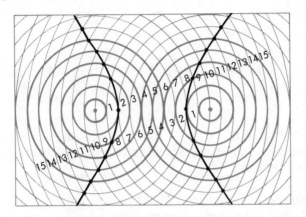

NOTES

1. These constructions yield the desired conics because they are based on the following loci definitions:

   Parabola—the set of points such that each is equidistant from a fixed point and a fixed line.

   Ellipse—the set of points such that the sum of the distances of each from two fixed points is constant.

   Hyperbola—the set of points such that the differences of the distances of each from two fixed points is constant.

2. A family of conics can be drawn on each worksheet. Just measure from a different line than zero for more parabolas. Let the sum and differences be some values other than 12 and 6 for more ellipses and hyperbolas.

3. Transparencies of these worksheets can be helpful in discussing the constructions.

EXPERIMENT 3: Paper folding the conics

OBJECTIVE

To show how the conics can be formed as envelopes of tangents by paper folding.

MATERIAL

A sheet of paper with a line on it and two sheets with circles.

DIRECTIONS

On each sheet mark a point as shown. Repeatedly fold the point onto the line or circle until the conic becomes apparent. The more creases made, the smoother the curve.

Fold a point not on the line onto various points of the line. The creases form tangents to a parabola.

PARABOLA

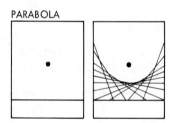

Fold a point inside the circle onto various points of the circle. The creases form tangents to an ellipse.

ELLIPSE

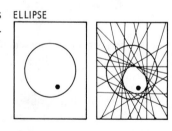

Fold a point outside the circle onto various points of the circle. The creases form tangents to a hyperbola.

HYPERBOLA

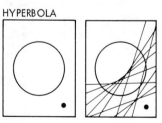

NOTES

1. Excellent examples of these curves can be formed using waxed paper where the creases are more vivid. These can then be easily projected on the overhead projector.
2. Geometry students should be able to prove that these constructions give the conics.

Point P is folded onto the line at P'. The crease is tangent to the parabola at point A. The focal point of the parabola is P.

The crease is the perpendicular bisector of segment PP', so PA and $P'A$ are equal in length. Since $P'A$ is perpendicular to the given line, it measures the distance from A to that line. Therefore, point A is equidistant from point P and the line and must lie on the parabola.

PARABOLA

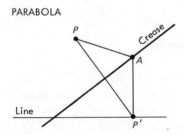

Point P is folded onto circle O at P'. The crease is tangent to the ellipse at point A. The foci of the ellipse are at P and O.

The crease is the perpendicular bisector of segment PP', so PA and $P'A$ are equal in length. Hence PA plus AO are equal in length to $P'A$ plus AO, which is the radius. Since the radius is constant in length, the sum of the distances of A from P and O is constant. Thus point A must be on the ellipse.

ELLIPSE

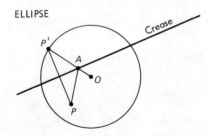

Point P is folded onto circle O at P'. The crease is tangent to the hyperbola at point A. The foci of the hyperbola are at P and O.

The crease is the perpendicular bisector of segment *PP′*, so *PA* and *P′A* are equal in length. Hence *PA* less *AO* is equal in length to *P′A* less *AO,* or simply *OP′*. But *OP′* is a radius and constant in length. Therefore, the difference of the distances of *A* from *P* and *O* is constant. Thus point *A* must be on the hyperbola.

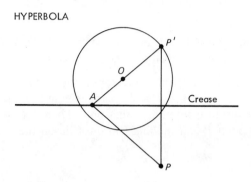

HYPERBOLA

4.10   Using nomographs

The word *nomograph,* which comes from the Greek and means a written law, applies in mathematics to a graphic technique for computing and for the solution of specific equations. Nomographs can serve as the source of many experiments that lead to discoveries and reviewing skills. Students gain valuable experience in constructing these graphs and in reading their scales as well as in verifying the results of computational problems.

Various examples of nomographs are given here together with suggested uses in the classroom. In each case, the greatest value comes from a student's constructing and reading his own nomograph, hence their inclusion here as laboratory experiments for the students. Additional examples of possible uses of nomographs in the mathematics class can be found in these publications:

DENHOLM, RICHARD. *Making and Using Graphs and Nomographs,* teacher's ed. Pasadena, Calif.: Franklin Publications, 1968.

SOBEL, MAX A., and EVAN M. MALETSKY. *Essentials of Mathematics,* Book 4. Lexington, Mass.: Ginn and Company, 1973.

The simplest example is the parallel scales nomograph. Better students can construct them from scratch using a ruler. With slower students you may prefer that they be constructed on graph paper.

These examples of nomographs involve addition in the form $a + b = c$. The scales are parallel with equal spaces between them. Outer scales are calibrated alike while the middle scale is reduced to half-size.

Adding integers

SUGGESTIONS FOR USE

1. To find the sum of two integers. [See example: $8 + (-3) = 5$.]
2. To illustrate that each integer and its opposite add to 0.
3. To show that the sum of two positive integers is positive and the sum of two negative integers is negative.
4. To show that the sign of the sum of two integers of opposite sign is always that of the integer farther from 0.
5. To illustrate the commutative property for addition of integers.

EXTENSIONS

6. To find the difference of two integers. [See example: $5 - (-3) = 8$.] Recall that each addition problem, $a + b = c$, can be written as a subtraction problem, $c - b = a$.
7. To show that the subtraction of an integer is equivalent to adding its opposite.

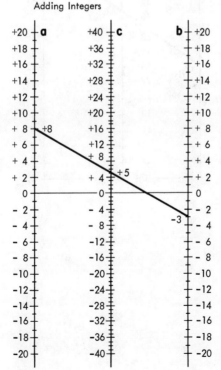

Adding Integers

Base five addition

Suggestions for Use

1. To develop skill in counting in base five as needed for calibration of scales.
2. To find the sum of two numbers expressed in base five notation. [See example: $12_{\text{five}} + 24_{\text{five}} = 41_{\text{five}}$.]

Extension

3. To find the difference of two numbers expressed in base five notation. [See example: $41_{\text{five}} - 24_{\text{five}} = 12_{\text{five}}$.]

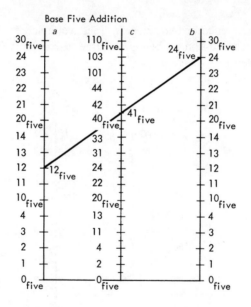

Base Five Addition

Nomograph theory

The theory behind the construction of parallel scales nomographs can be explored by your better students.

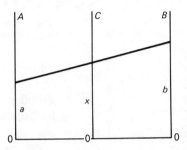

The segment x is the median of the trapezoid with parallel bases a and b. Its length is half that of the sum of the lengths of the bases. By calibrating the C scale to half that of the outer scales A and B, it can be used to read the sum directly.

A slight modification gives a nomograph for the addition $a + 2b = c$. Just calibrate the B scale at twice that of the A scale, keeping the C scale at half the A scale. Students can be asked to calibrate a nomograph this way and try to discover the relationship among a, b, and c. A nomograph of this type can be used to find the perimeter of an isosceles triangle where a measures the base and b the two equal sides.

Multiplication

One useful extension of the parallel scales nomograph is for multiplication. When the A, B, and C scales are logarithmic, the resulting effect is multiplication $(a \cdot b = c)$. Logarithms can be thought of as exponents and exponents are added in multiplication $(10^a \cdot 10^b = 10^{a+b} = 10^c)$. So, in effect, the nomograph still only adds.

Per cents

The nomograph shown here can be especially useful in reviewing percentages. Supply the student with the nomograph on a worksheet. It can be easily used without knowledge of logs or the log scale.

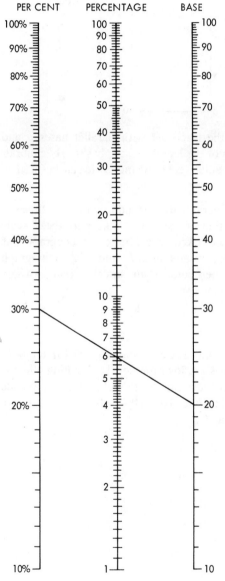

PER CENT   PERCENTAGE   BASE

SUGGESTIONS FOR USE

1. To offer experience in reading scales.

2. To offer a mechanical method for approximating solutions to per cent problems either to verify direct computation or to be verified by direct computation.

3. To find P given R and B, R given P and B, and B given P and R. (See example: 30% of 20 is 6.)

EXTENSION

Supply similar worksheets but with the scales not labeled. Then have your students label them so that they can be used to solve problems of this type.

4. $d = r \cdot t$

5. $A = b \cdot h$

Non-parallel-scale nomograph

Not all nomographs have parallel scales. This one can be used to find *k* given *a* and *b* in the relationship $1/a + 1/b = 1/k$.

SUGGESTIONS FOR USE

1. To find the sum of the reciprocals of two numbers.
   (See example: $\frac{1}{3} + \frac{1}{6} = \frac{1}{2}$.)

2. To offer a mechanical method of solving algebra problems such as:

   > Pipe *A* fills a tank in 3 hours while pipe *B* fills the same tank in 6 hours. How long will it take for both pipes to fill the tank together?   (Answer: 2 hours.)

3. To find the sum of two resistances in parallel:

   > A 6-ohm and a 3-ohm resistor are wired in parallel. What is the single equivalent resistance?   (Answer: 2 ohms.)

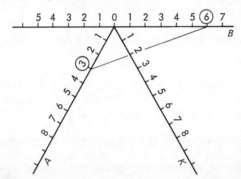

For more information on this particular type of nomograph, see "Notes on a 'Spider' Nomograph," by Lowell Van Tassel, *Mathematics Teacher,* November 1959.

Exercises

1. Find the largest number required from each sequence in forming an *n*-by-*n* magic square.
   (a) 1, 2, 3, 4, 5, . . . , ?
   (b) −10, −9, −8, −7, −6,   . . . , ?
   (c) 3, 6, 9, 12, 15, . . . , ?
   (d) −7, −4, −1, 2, 5, . . . , ?

2. Draw the 11 distinctly different patterns that can be used to form a model of a cube.

3. Verify that in an nth-order magic square formed from successive integers starting with 1,

   (a) the sum of all entries is $\dfrac{n^2(n^2+1)}{2}$.

   (b) the sum of each row, column, and diagonal is $\dfrac{n(n^2+1)}{2}$.

   (c) the middle entry is $\dfrac{n^2+1}{2}$.

4. Study the proof of Pick's formula, $A = b/2 + i - 1$, as given in *Mathematical Snapshots,* by Hugo Steinhaus (New York: Oxford University Press, Inc., 1969).

5. Twelve congruent square pieces of paper are cut out, six red and six green. How many different-colored models of a cube can be formed from any six of the squares?

6. Show how Euler's formula, $V + F = E + 2$, applies to a prism and a pyramid with an n-gon as a base.

7. Find the number of vertices, faces, and edges in each of the regular polyhedrons.

8. A diagonal is drawn on each face of a cube. Does Euler's formula, $V + A = R + 2$, apply to the network of vertices, arcs, and regions formed on the surface of the cube?

9. Prove that any network must have an even number of odd vertices.

10. Draw sketches of the first five hexagonal numbers starting with 1. What is the second differences for these numbers? What is the nth hexagonal number?

11. Show how to paperfold a regular hexagon and octagon.

12. Form a 16-card deck from 4 jacks, 4 queens, 4 kings, and 4 aces. Repeatedly shuffle the deck, dealing out 20 three-card hands. Count the number of hands that have at least two cards of the same value and compare the results with the theoretical computed probability value.

13. Repeat Experiment 1, page 130, on probability.

14. Prove that the sum of the first n odd counting numbers is n^2.

15. Prove that the "spider" nomograph on page 147 produces the relationship $1/a + 1/b = 1/k$.

Activities

1. Read and report on Chapter 7, "Aids and Activities," by Evan M. Maletsky in the Thirty-fifth Yearbook of the National Council of Teachers of Mathematics, *The Slow Learner in Mathematics* (Washington, D.C.: N.C.T.M., 1972).

2. Construct a laboratory worksheet that can be used to lead a student to discover that the sum of each row, column, and diagonal in an nth-order magic square formed from the integers 1 through n^2 is $\dfrac{n(n^2 + 1)}{2}$.

3. Construct a magic square on a model of a torus as described in Martin Gardner's *Second Scientific America Book of Puzzles and Diversions* (New York: Simon and Schuster, 1961).

4. Research on the interest in magic squares shown by the statesman Benjamin Franklin and the artist Albrecht Dürer.

5. Prepare a set of activity cards that a student could use with a geoboard to develop a specific geometric formula or property.

6. Prepare a set of student activity cards that could be used to discover and develop properties in Pascal's triangle.

7. Construct a model of a 3 x 3 x 3 magic cube as described on page 106.

8. Prepare a lesson plan on Euler's formula for an eighth-grade class. Pay specific attention to the motivation of the lesson, the use of classroom models, individual student involvement, and an appropriate assignment.

9. Form some solids from a set of congruent cubes by joining them face to face. Does Euler's formula appear to apply to the surface of these solids when counting vertices, faces, and edges of the individual cubes? Where vertices and edges coincide, count them only once.

10. Develop an experiment that students could follow in discovering the pattern relating the number of diagonals to the number of vertices in a polygon.

11. Develop a paper-folding experiment that can be used to illustrate how to find the area of a parallelogram. Use the same method to illustrate that the diagonals of a parallelogram bisect each other and that the

midpoint of the hypotenuse of a right triangle is equidistant from the three vertices.

12. See how many different circle relationships can be easily demonstrated using paper-folding illustrations.

13. Experiment with Buffon's needle problem using at least 200 drops.

14. Ask 23 people selected at random and see if you find at least one duplication of birthdates.

15. Prepare a lesson introducing probability to a first-year general math class.

16. Investigate the uses of probability by insurance companies in establishing their mortality tables.

17. Follow Experiment 3 in Section 4.9 to fold a parabola, ellipse, and hyperbola on waxed paper that can be shown on the overhead projector.

18. Prepare a lesson on adding integers that makes use of student-constructed nomographs.

19. Construct a per cent nomograph from strips of one- and two-cycle semilog graph paper.

20. Discuss how a nomograph for finding unit prices can be used in teaching comparison shopping in a general mathematics class.

References and selected readings

ANDREWS, W. S. *Magic Squares and Cubes.* New York: Dover Publications, Inc., 1960.

BECK, ANATOLE, M. N. BLEICHER, and D. W. CROWE. *Excursions into Mathematics.* New York: Worth Publishers, Inc., 1969, Chap. 1, "Euler's Formula for Polyhedra, and Related Topics."

CUNDY, H. MARTYN, and A. P. ROLLETT. *Mathematical Models.* New York: Oxford University Press, Inc., 1961.

DENHOLM, RICHARD. *Making and Using Graphs and Nomographs,* teacher's ed. Pasadena, Calif.: Franklin Publications, 1968.

FITZGERALD, W., D. BELLAMY, P. BOONSTRA, J. JONES, and W. OOSSE. *Laboratory Manual of Elementary Mathematics.* Boston: Prindle, Weber & Schmidt, Inc., 1969.

JOHNSON, DONOVAN A. *Paper Folding for the Mathematics Class*. Washington, D.C.: National Council of Teachers of Mathematics, 1957.

JOHNSON, DONOVAN A., and G. R. RISING. *Guidelines for Teaching Mathematics*. Belmont, Calif.: Wadsworth Publishing Company, Inc., 1972, Chap. 27, "The Mathematics Laboratory."

KIDD, KENNETH, S. S. MYERS, and D. M. CILLEY. *The Laboratory Approach to Mathematics*. Chicago: Science Research Associates, Inc., 1970.

KRULIK, STEPHEN. *A Handbook of Aids for Teaching Junior–Senior High School Mathematics*. Philadelphia: W. B. Saunders Company, 1970.

———. *A Mathematics Laboratory Handbook for Secondary School*. Philadelphia: W. B. Saunders Company, 1972.

MADACHY, JOSEPH S. *Mathematics on Vacation*. New York: Charles Scribner's Sons, 1966, Chap. 3, 4.

Mathematics Teacher. "Conic Sections in Relation to Physics and Astronomy" (Feb. 1970), p. 101.

———. "Conics from Straight Lines and Circles" (Mar. 1973), p. 243.

———. "Enrichment: A Geometry Laboratory" (Mar. 1963), p. 134.

———. "Grading Nomograph" (Nov. 1965), p. 595.

———. "If Pythagoras Had a Geoboard" (Mar. 1973), p. 215.

———. "The Limit Concept on a Geoboard" (Jan. 1972), p. 13.

———. "The Magic of a Square." (Jan. 1970), p. 5.

———. "Manipulating Magic Squares" (Dec. 1972), p. 729.

———. "The Mathematics Laboratory" (Mar. 1963), p. 141.

———. "Nomography" (Nov. 1961), p. 531.

———. "Notes on a 'Spider' Nomograph" (Nov. 1959), p. 557.

———. "Organizing a Mathematics Laboratory" (Feb. 1967), p. 117.

———. "A Psychedelic Approach to Conic Sections" (Dec. 1970), p. 657.

———. "Straight Line Model for Multiplication" (Apr. 1966), p. 343.

National Council of Teachers of Mathematics. *Experiences in Mathematical Discovery,* Washington, D.C.: N.C.T.M., 1966, 1967, 1970, 1971. Vols. 1–7, 9.

———. *The Slow Learner in Mathematics* (Thirty-fifth Yearbook). Washington, D.C.: N.C.T.M., 1972, Chap. 8, "Laboratory Approach."

PEARCY, J. F. F., and K. LEWIS. *Experiences in Mathematics, Stage 1, Stage 2, and Stage 3*. Boston: Houghton Mifflin Company, 1966, 1967.

Row, T. Sundara. *Geometric Exercises in Paper Folding.* New York: Dover Publications, Inc., 1966.

Sobel, Max A. *Teaching General Mathematics.* Englewood Cliffs, N.J.: Prentice-Hall, Inc., 1967.

Steinhaus, Hugo. *Mathematical Snapshots.* New York: Oxford University Press, Inc., 1969.

Wenninger, Magnus J. *Polyhedron Models for the Classroom.* Washington, D.C.: National Council of Teachers of Mathematics, 1966.

CHAPTER FIVE

CLASSROOM AIDS AND MODELS

The art of teaching is the art of communicating interest and ideas, and the teacher who can express and illustrate these in a variety of ways stands the best chance of getting them across. This chapter is devoted to examples of aids and models that can be constructed by the teacher or the student to serve as a visual means of communicating an idea. Some can be used in introducing new ideas and others in reviewing old ones. Some have an inherent novelty about them that will help attract attention while others present familiar material from a fresh approach. Some help review the skills while others help tax and challenge the imagination. But they all give a new visual dimension to a mathematical idea.

The chapter is divided into four parts: arithmetic skills, informal geometry, geometric models, and algebraic concepts. Detailed construction steps are given for aids and models in each area along with suggested uses and some possible extensions. Beyond those described herein, many of the experiments, activities, and ideas in earlier chapters are also adaptable to this format and use.

5.1 Aids for maintaining arithmetic skills

Once the various arithmetic skills have been introduced and developed for the student, they need to be maintained throughout the rest of his mathematical training. This can often be a very difficult task since it requires a new look, a fresh approach, to a familiar topic. Reviewing and drilling on

154

the skills in the same way that they were introduced is to the student un-imaginative, unmotivating, and all too often unsuccessful.

Visual aids can frequently serve as the vehicle for review, as a new disguise for an otherwise recognizable subject. Consider the teacher who wants to spend a few minutes at the beginning of class reviewing whole-number computation. Start with a number on the board, today's date but not in the usual form. Here's an example:

21475

(2/14/75 for February 14, 1975)

The first task is to get the class to recognize the number. Next the teacher brings out a set of cards numbered with these digits. The cards are mixed up and a student selects one. Then each student or team sees if they can be the first to use the other four cards with the fundamental operations to get the number selected. If the same card is selected a second time, a new answer must be given. Here are some possible answers for 21475, assuming that 7 is selected as the objective card.

$$5 + 4 - (2 \times 1) = 7$$
$$(2 \times 5) + 1 - 4 = 7$$
$$(5 \times 1) + (4 \div 2) = 7$$

Numbered cards of this type can have many uses in reviewing arithmetic skills. It is a simple idea, a simple aid, a simple activity—but far more motivating than a drill sheet of problems.

Napier's rods

Invented by the Scottish nobleman John Napier (1550–1617), these simple computing devices were in common use during much of the 1600s. They were designed to simplify the burdensome task of multiplication by the very man who later invented logarithms—which, in effect, made multi-plication problems addition problems. The original rods were made from strips of wood or bone and small enough to carry in the pocket. Each rod had four sides with a scale on each side. By placing the appropriate rods side by side, you had a convenient computing device for multiplying quickly.

Napier's rods can be an interesting and exciting topic for students at most levels of ability, especially when placed in their proper historical per-spective. Several methods of constructing demonstration models are given here. This activity offers an excellent opportunity for individual student involvement.

CONSTRUCTION

For the chalkboard, mark a large sheet of poster paper as shown, using a special color for the index numbers. Cut it into strips and set them on the chalk ledge or attach them to the board for demonstration.

For the overhead projector, copy them on a sheet of acetate, cut out the strips, and project them on the screen or board.

Index	1	2	3	4	5	6	7	8	9	0
1	0 / 1	0 / 2	0 / 3	0 / 4	0 / 5	0 / 6	0 / 7	0 / 8	0 / 9	0 / 0
2	0 / 2	0 / 4	0 / 6	0 / 8	1 / 0	1 / 2	1 / 4	1 / 6	1 / 8	0 / 0
3	0 / 3	0 / 6	0 / 9	1 / 2	1 / 5	1 / 8	2 / 1	2 / 4	2 /	0 / 0
4	0 / 4	0 / 8	1 / 2	1 / 6	2 / 0	2 / 4	2 / 8	3 / 2	3 / 6	0 / 0
5	0 / 5	1 / 0	1 / 5	2 / 0	2 / 5	3 / 0	3 / 5	4 / 0	4 / 5	0 / 0
6	0 / 6	1 / 2	1 / 8	2 / 4	3 / 0	3 / 6	4 / 2	4 / 8	5 / 4	0 / 0
7	0 / 7	1 / 4	2 / 1	2 / 8	3 / 5	4 / 2	4 / 9	5 / 6	6 / 3	0 / 0
8	0 / 8	1 / 6	2 / 4	3 / 2	4 / 0	4 / 8	5 / 6	6 / 4	7 / 2	0 / 0
9	0 / 9	1 / 8	2 / 7	3 / 6	4 / 5	5 / 4	6 / 3	7 / 2	8 / 1	0 / 0

USES

1. Place the index rod along side one of the others and show how the corresponding multiples are given. For example, with the 8-rod and the index, ask for such products as 7×8.

2. Repeat using two- and three-digit factors. For example, show 7 × 86 and 7 × 862. Explain how "carrying" works when adding along the diagonals. Let students read other products involving 8, 86, and 862 from the same settings.

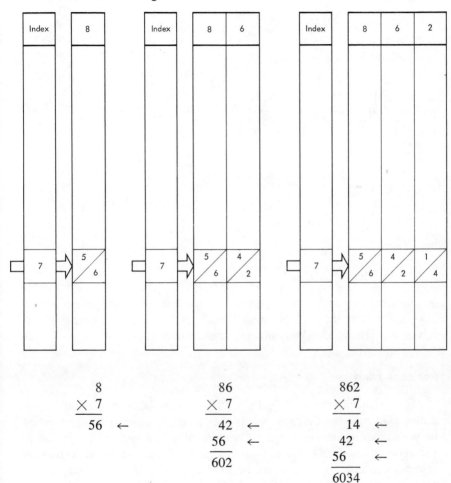

$$\begin{array}{r} 8 \\ \times\ 7 \\ \hline 56 \ \leftarrow \end{array}$$

$$\begin{array}{r} 86 \\ \times\ 7 \\ \hline 42 \ \leftarrow \\ 56 \ \leftarrow \\ \hline 602 \end{array}$$

$$\begin{array}{r} 862 \\ \times\ 7 \\ \hline 14 \ \leftarrow \\ 42 \ \leftarrow \\ 56 \ \leftarrow \\ \hline 6034 \end{array}$$

3. Allow students to suggest other multiplication problems, still with one-digit multipliers, and let others illustrate them at the board. See that some include the 0-rod.

4. Relate the results shown on the rods to the common computing algorithm now used for multiplying.

5. Let students discover how the rods can be used with two- and three-digit multipliers.

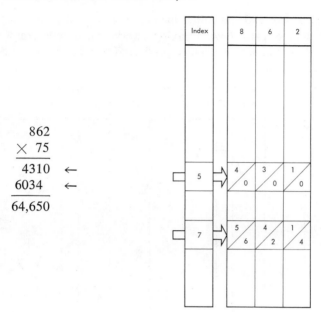

$$
\begin{array}{r}
862 \\
\times\ 75 \\
\hline
4310 \leftarrow \\
6034 \leftarrow \\
\hline
64,650
\end{array}
$$

SUGGESTED EXTENSION

Students can make their own set using strips of ½-inch graph paper. Let them review the multiplication facts by filling in all the entries themselves and also by checking answers read from the rods.

Per cent chart

Many students have difficulty with per cents both in the junior and senior high schools. Perhaps part of this problem arises from the method in which per cents are taught. Concrete aids and activities relating to per cents often clarify and reinforce what for some can be an extremely abstract concept. This chart can be used effectively in class as an aid for actually finding per cents that are then to be verified by computation.

CONSTRUCTION

Rule the grid shown on a large piece of poster paper. Attach a string through a hole at the top of the per cent scale. Make the string long enough so that it can be stretched tightly to any point on the percentage and base scale. Mount the entire chart on the bulletin board or chalkboard. By properly positioning the string, the answers to the various forms of per cent problems can be read directly from the chart.

USES

1. Find a percentage given the rate and base.

What is 60% of 50?

Rate   60%

Base   50

Percentage   ?

Stretch the string through a base of 50. Read across at the per cent 60. The percentage is 30.

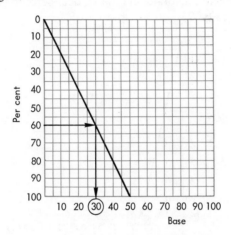

2. Find a per cent given percentage and base.

What per cent of 75 is 60?

Base   75

Percentage   60

Per cent   ?

Stretch the string through a base of 75. Read up at a percentage of 60. Reading across, the per cent is found to be 80.

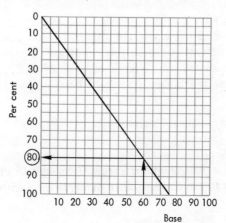

3. Find the base given the per cent and percentage.

40% of what number is 10?

Per cent   40
Percentage   10
Base   ?

Stretch the string such that it passes through the intersection of the horizontal per cent line of 40 and the vertical percentage line of 10. The base is found to be 25.

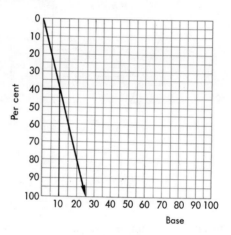

Base

SUGGESTED EXTENSION

Students can construct their own per cent charts on graph paper using a ruler instead of a string to read the results.

Binary cards

Here is an interesting classroom excursion that can be used as the culmination of a unit of study on other systems of numeration. To help answer students' questions of "What good is all of this?", this activity purports to show how the binary system of notation may be used in card sorting.

CONSTRUCTION

As a first step, write the numbers from 0 through 15 in binary notation. Prefix zeros so as to have a four-digit numeral for each number.

Base 10	Base 2	Base 10	Base 2
0	0000	8	1000
1	0001	9	1001
2	0010	10	1010
3	0011	11	1011
4	0100	12	1100
5	0101	13	1101
6	0110	14	1110
7	0111	15	1111

Next each student must prepare a set of 16 index cards, with four holes punched in each and with one of the corners cut off for purposes of identifying the face of the card. Now the numbers from 0 through 15 are represented as in the following figures. In each case a hole alone represents 0, while a slot cut above the hole to the edge of the card represents 1. The number that each card represents should be written on the face of the card, as shown.

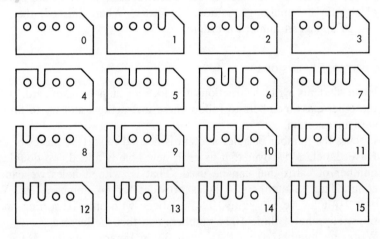

USES

Here are several card-sorting experiments that can now be performed with these cards:

1. Place the cards in a stack and shuffle them well. How can all the cards that represent even numbers be located? We note that the binary notation for all even numbers has a 0 in the unit's place. Therefore, use a pencil, paper clip, or other sharp object and push through the first hole from the right. Lift up and you will have located all the even numbers.

2. With only four operations, any number from 0 through 15 can be located. For example, let us direct a card-sorting machine to locate the card that represents 13. Working from right to left, these are the directions:

Punch through, discard those pulled.
Punch through, keep only those pulled.
Punch through, discard those pulled.
Punch through, discard those pulled.

$13_{ten} = 1101_{two}$

Note that in any position when you punch through you obtain all the cards that have a 0 in that binary place. Thus, you keep the cards if the number in question has a 0 in a particular position; otherwise the cards are discarded.

3. With only four operations, the shuffled cards can be placed back in numerical order from 0 through 15. Punch through the first hole from the right. Place all the cards that lift up (those with a 0 in the unit's place) in front of the other cards. Then repeat this procedure for each of the three remaining positions, in order. When finished, the cards should appear in numerical order.

SUGGESTED EXTENSIONS

1. Have students write instructions for locating a particular card. Allow the other class members to follow these instructions and try to find the card in question.

2. Discuss the number of holes and cards needed to sort 32 pieces of data. Have the class discover that as each new hole is added, we double the number of cards that can be used. That is, with 5 holes we can accommodate 32 numbers, 6 holes provides for 64 numbers, etc. With only 10 holes we can write 1024 numbers; with only 10 sorting operations, any one of these 1024 numbers can then be located.

3. Write the numbers from 0 through 31 in binary notation. Have students prepare a set of 32 cards, each with 5 holes, to represent these numbers. Then use these cards to perform experiments similar to those described earlier.

REFERENCES

SOBEL, MAX A. *Teaching General Mathematics.* Englewood Cliffs, N.J.: Prentice-Hall, Inc., 1967, p. 24.

A set of commercially made demonstration cards may be obtained from Midwest Publications Company, Inc., P.O. Box 307, Birmingham, Mich. 48012.

5.2 Aids for informal geometry

Most of the informal geometric ideas presented in the mathematics class lend themselves quite naturally to the use of visual aids, but there are always additional ideas that can be collected and tried as well. For example, when studying right triangles, it is interesting to show how the ancient Egyptian rope-stretchers might have used a piece of knotted rope to form a right angle. Not only will students be exposed to the Pythagorean property, but they will also have an appreciation of the importance of this famous theorem to the ancient Egyptians, who, are said to have used it to resurvey their land following the annual flooding of the Nile River several thousands of years before Pythagoras was born.

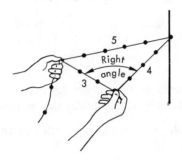

Pythagorean property

Many different aids can be used in teaching the Pythagorean property. Some serve best for initial investigation and discovery while others are best in review. Some are very formal and exact, others are informal and approximate. Some are suitable only with a 3-4-5 right triangle.

This aid is best suited for review of the Pythagorean property. It is simple to make and use by the teacher and dynamic and visual to the viewer. Although it is not a proof, it does leave in the eyes of the observer a convincing result for any right triangle.

CONSTRUCTION

Cut a right triangle of any size from red paper. Then cut a large square of green paper to fit the hypotenuse and two smaller squares of white paper to fit the two legs. Tape the squares onto the sides of the triangle as shown here.

Uses

1. The object is to show a simple method of cutting the squares on the legs so that the pieces can be easily rearranged to fit the square on the hypotenuse.

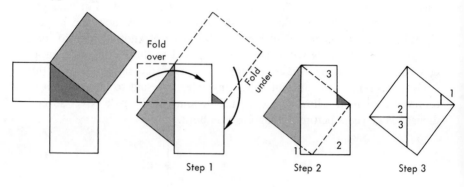

Step 1   Step 2   Step 3

Step 1: Fold the square on the smaller leg over the triangle. Fold the square on the hypotenuse under the triangle.

Step 2: Cut off the three triangular pieces that extend beyond the large square as shown.

Step 3: Relocate these pieces as shown to completely cover the square on the hypotenuse.

By using the colors suggested, the white areas (that of the squares on the two legs) will completely cover the green area (that of the square on the hypotenuse), thereby illustrating the theorem.

2. Repeat with several right triangles of other sizes. Then let the students construct some of their own to illustrate the same propery. Follow the same steps; only the sizes of the pieces cut and rearranged will differ.

Suggested Extensions

Choose the lengths of the two legs of a right triangle. Next have students lay out a tiling pattern with squares of these two sizes as shown. Graph paper may help. Then see if they can draw on it a square with sides corresponding in size to that of the hypotenuse and rearrange the pieces of the two tiling squares to fit. The figure formed is exactly that of the aid described above.

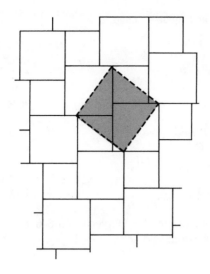

Prisms and pyramids

Students are often asked to construct or solve problems involving particular polyhedrons with given dimensions. This aid helps in the development of space perception by having students search for and build various prisms and pyramids.

CONSTRUCTION

Cut the indicated number of each piece from construction paper of a different color.

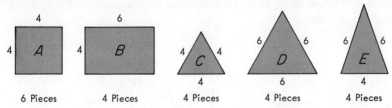

USES

Have individual students demonstrate how various prisms and pyramids can be formed from selected pieces. Form as many different ones as possible.

165

Rectangular Prisms	*Rectangular Pyramids*
AAAAAA	ACCCC
AABBBB	AEEEE
	BDDEE
	BCEEE

Triangular Prisms	*Triangular Pyramids*
AAACC	CCCC
ABBEE	DDDD
BBBCC	CEEE
BBBDD	DDEE

For a simpler version, delete the D pieces (6-6-6 equilateral triangles). For a more difficult version, add some new F pieces (4-4-6 triangles).

SUGGESTED EXTENSION

An exciting card game related to forming polyhedrons is *Polyhedron-Rummy,* available from Scott, Foresman and Company, 1900 East Lake Ave., Glenview, Ill. 60025.

Cones

A very simple aid, a circular piece of paper, can be used to introduce the student not only to the cone but to graphing, variation, and the limit concept.

CONSTRUCTION

Cut a circular piece of paper about 10 inches in diameter and slit it through to the center.

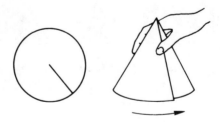

USES

1. Form several cones from the paper and in each case describe the height, slant height, and radius. These three measures of the cone are often confused by the student.

2. Illustrate how the cone changes in size and shape by curling up the paper model as shown here.

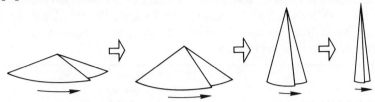

Have the students describe in their own words how each of these changes as the cone changes shape.

height	area of base
slant height	area of cone
radius	volume of cone

3. The elements of the limit concept should become part of the student's thinking long before he sees it in the context of the calculus. Encourage some informal thinking along those lines with these questions.

As the height approaches 0,
what length does the radius approach?   (slant height)
what does the volume approach?   (0)
what does the surface area approach?   (twice area of original circle)

As the length of the radius approaches 0,
what does the height approach?   (slant height)
what does the volume approach?   (0)
what does the surface area approach?   (0)

4. Set up appropriate axes and see how the students graph the relationship developed in steps 2 and 3.

Tetratetraflexagons

This amazing little geometric model appears to have just two sides, front and back. But by flexing the model, both sides can be made to disappear. It is called a *tetratetraflexagon* because it can be flexed to show four (tetra) different sides or faces and because it has a four (tetra)-edged rectangular shape.

Constructing and assembling the tetratetraflexagon requires care, patience, and a certain degree of skill, but the effort will be more than repaid by the novelty of the completed model. Students will want their own model to keep.

CONSTRUCTION

1. Reproduce the pattern and markings as shown so that each student can get a copy. Starting with small 2-inch squares, the initial pattern will be 6 by 8. The size may be changed, but it is advisable to avoid extremely large or small models.

		DIRECTIONS: Fold vertically down the center so the two halves are back to back. Carefully separate the two parts at the fold.	
TETRA- **TETRA-**	**FLEXAGON**		
	A new side should appear. Flex it again and you'll find still another side. Remember, don't tear the paper.		You may think this has just two sides, front and back. If you do, you're wrong!
SIDE 1	When you flex this paper in just the right way, you can make the writing disappear.	Now have fun trying to flex them back! **SIDE 2**	

2. Cut out the pattern. Then repeatedly fold it back and forth along each of the three vertical lines. With a knife or a razor blade cut around three sides of the two center squares as shown. Be sure not to completely detach them from the rest.
3. Closely follow these assembling instructions.

Position with word tetratetraflexagon in upper left-hand corner.

Fold center flap behind left hand column and also fold back right hand column.

Fold back right hand column a second time.

Fold end of flap over on top and tape *only* on squares shown.

4. Follow directions on the tetratetraflexagon for flexing. Be careful that you do not force a fold or tear the paper. By flexing the model twice, both faces with the writing should disappear.

SUGGESTED EXTENSIONS

If this flexagon proves interesting, try assembling and flexing a hexa-hexaflexagon—six faces formed from a six-sided polygon. A pattern is given in the first reference below.

REFERENCES

GARDNER, MARTIN, ed. *Scientific American Book of Mathematical Puzzles and Diversions.* New York: Simon and Schuster, 1959, Chap. 1, "Hexaflexagon," pp. 1–14.

————. *Second Scientific American Book of Mathematics Puzzles and Diversions.* New York: Simon and Schuster, 1961, Chap. 2, "Tetraflexagons," pp. 24–31.

Clinometer and hypsometer

Many students enjoy field activities in the mathematics class. The clinometer and hyposometer are simple, useful instruments designed to help determine the height of an object by indirect measurement. They can be made and used by the students themselves with results that can be surprisingly accurate. Adaptable to a variety of levels, they can be effectively used as a supplement to units on measurement, scale drawing, and ratios, or even as an introduction to numerical trigonometry.

The principle of the clinometer and hypsometer is based on similar figures.

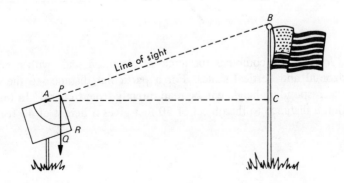

Triangle *PQR* is similar to triangle *ABC*. The plumb line determines the vertical line *PQ;* the horizontal line *AC* is determined by sighting with the plumb line aligned to the edge of the hypsometer *PR*. The distance from *C* to the ground (the height of the instrument on level ground) must be added to the height read from the hypsometer to determine the actual height of the flagpole.

CONSTRUCTION

Students can draw the scales themselves or attach a copy supplied by the teacher on a rectangular piece of cardboard. Suspend a plumb line from the upper right-hand corner of the grid as shown and attach it to a pole of convenient height so that it will pivot freely.

1. The simplest version is a *clinometer,* which is used to measure angles of elevation. The size of the angle is read directly from the scale at the point where the plumb line crosses it. In the illustration this angle is 40°.

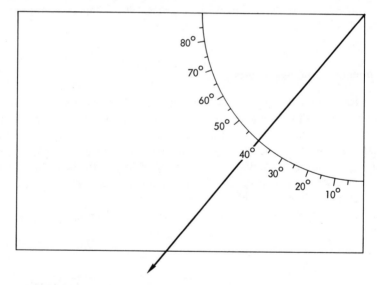

2. The *hypsometer* combines the angle of elevation scale with a grid and horizontal and vertical scales. For a particular distance to the object, the corresponding height can be read directly from the grid. In this illustration a distance to the object of 50 feet gives a height of 42 feet.

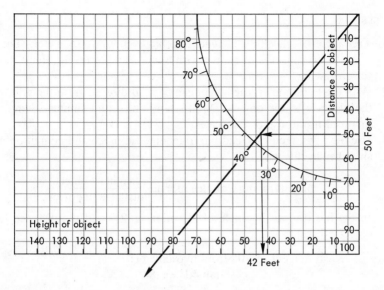

3. By drawing an arc of unit (in this case 100) radius, the use of the hypsometer can be extended to include reading the sine, cosine, and tangent of the angle of elevation.

$$\sin 40° = 0.64$$
$$\cos 40° = 0.77$$
$$\tan 40° = 0.84$$

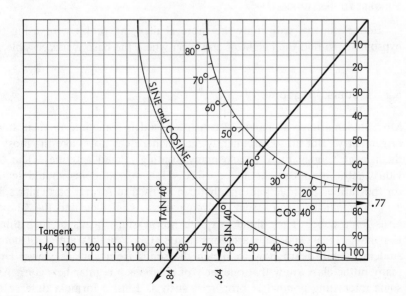

Uses

1. The angle of elevation is read directly from the clinometer. The distance to the object is measured with a tape and the height found through scale drawing.
2. With the hypsometer, the height can be read directly from the grid, given the distance to the object. Discuss how ratios and similar triangles are involved in the use of the scales.
3. Use the hypsometer to read the trigonometric ratios and compute the desired height this way.

$$\text{tangent of angle of elevation} = \frac{\text{height of object}}{\text{distance to object}}$$

In the illustration the angle of elevation is 40° and the distance to the object 50 feet.

Read tangent of 40° from hypsometer:
$$\tan 40° = .84$$
Compute height from ratio above:
$$.84 = \frac{h}{50}$$
$$h = 42$$

The height is 42 feet plus height of instrument.

Suggested Extension

Better students will enjoy the more challenging problem of using the hypsometer to find the height of an object where the base is inaccessible.

5.3 Models in geometry

Most students have a chance to construct models sometime in their study of geometry. But all too frequently the teacher does not make the most of them, even though there are so many exciting related activities. The obvious metric questions should be asked. However, students should search for the various possible patterns that can be used to form a particular solid as well as construct one from a given pattern. Can they find the 11 patterns for a cube? They should explore how various boxes of such things as candy, cereal, and nails are constructed. Once models are constructed, students should learn how to sketch the solids from various views. How many in the class know that one view of a cube is a regular hexagon? And some interesting nonmetric properties such as Euler's formula deserve attention as well.

Some excellent references on this topic are:

CUNDY, H. MARTYN, and A. P. ROLLETT. *Mathematics Models.* New York: Oxford University Press, Inc., 1961.

HOLDEN, ALAN. *Shapes, Space, and Symmetry.* New York: Columbia University Press, 1971.

Regular polyhedrons

The literature in the history of mathematics is rich with stories about regular polyhedrons. Their unique special beauty and symmetry has intrigued men in all ages of time. Some of the solids were probably known to the ancient Egyptians, and the Pythagoreans (ca. 500 B.C.) discovered others. They were later known as the Platonic solids when the Greek, Plato (ca. 400 B.C.), wrote of their role in the design of matter with this mystical association to the four so-called "elements" of the universe:

tetrahedron	fire
hexahedron (cube)	earth
octahedron	air
icosahedron	water
dodecahedron	universe

Euclid (ca. 300 B.C.) devoted the last of the 13 books in his famous *Elements* almost entirely to these solids. Included there was a proof that only five such solids exist.

Centuries later Kepler (1571–1630) wrote in his *Mystery of the Cosmos* that God had in view the five regular solids of geometry in designing the heavens. He postulated that the radii of the orbits of the planets were proportional in length to those of the six spheres formed from an initial sphere successively inscribed with the five regular solids and their own inscribed spheres. So taken was he with this grand design of the universe that he argued for some time that only six planets could possibly exist because only five regular solids existed. It should be remembered that at this time only six planets had been found.

Today we can see the shapes of a cube and tetrahedron in crystals of sodium chlorate, an octahedron in crystals of chrome alum, and a dodecahedron and icosahedron in the skeletons of radiolaria, microscopic sea animals. Strange as it may seem, these ancient Platonic solids do describe the shape of some matter.

CONSTRUCTION

Models of these solids can best be constructed from heavy paper or tag, joining the edges with tape or glued tabs. With slower students the completed pattern or net should be supplied such that the only activity is cutting and assembling. Better students enjoy copying the nets as well. One method is to cut a template in the shape of each basic face from a 3- by 5-inch file card. The net can then be laid out accurately by marking each vertex of each face with a pencil or pin.

The five regular polyhedrons have but three basic shapes for their faces: the equilateral triangle, the square, and the regular pentagon.

Regular polyhedron	Number of faces	Shape of faces
Tetrahedron	4	equilateral triangles
Hexahedron (cube)	6	squares
Octahedron	8	equilateral triangles
Dodecahedron	12	regular pentagons
Icosahedron	20	equilateral triangles

One set of possible patterns for construction is shown here.

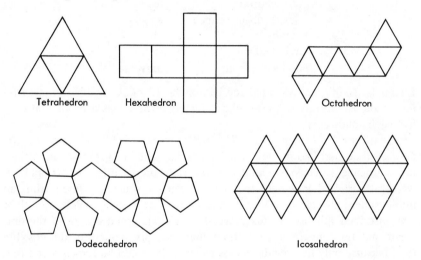

Tetrahedron   Hexahedron   Octahedron

Dodecahedron   Icosahedron

USES

The skills involved in the careful and accurate construction of these solids are often neglected in the mathematics class. As an activity, it can be justified for no other reason. However, work with these solids can help develop still another important yet often neglected skill, that of three-dimensional visualization.

Here are several suggestions that make use of these solids in developing space awareness and perception.

1. Using pencils as lines located by edges and 3- by 5-inch file cards as planes located by faces, describe possible relationships among lines and planes in space, such as parallelism, perpendicularity, and skewness.

   For example, use one face of a solid to locate a plane. Then see how many parallel planes can be located from other faces. Except for the tetrahedron, each solid contains pairs of parallel faces.

   As another example, locate a line using an edge of a solid. Then see how many parallel lines can be located from other edges. Except for the tetrahedron, each solid contains pairs of parallel edges, although they may not always be easy to find.

2. Not many students, looking at solids such as these, are aware of their various parts. Ask the students to count vertices, faces, and edges. See who can come up with simple counting techniques.

   Edges are joined at vertices and faces joined at edges. A dodecahedron has 12 pentagonal faces. The separate edges of the individual faces ($12 \times 5 = 60$) are paired to form the edges ($60/2 = 30$) of the solid. The separate vertices ($12 \times 5 = 60$) are joined three at a time for the ($60/3 = 20$) vertices of the solid.

Regular polyhedron	Number of vertices	faces	edges
Tetrahedron	4	4	6
Hexahedron (cube)	8	6	12
Octahedron	6	8	12
Dodecahedron	20	12	30
Icosahedron	12	20	30

Each of these number triples satisfies Euler's formula, $V + F = E + 2$.

3. An interesting and challenging question related to the regular polyhedrons involves coloring. Faces of each solid are painted, each entirely with a single color. What is the least number of colors needed to paint each solid such that no two adjacent faces have the same color? The answers are surprising!

REFERENCES

HOLDEN, ALAN. *Shapes, Space, and Symmetry.* New York: Columbia University Press, 1971.

WENNINGER, MAGNUS J. *Polyhedron Models for the Classroom.* Washington, D.C.: National Council of Teachers of Mathematics, 1966.

Geometric models by plaiting

To make most models of polyhedrons rigid, the edges must be taped or extra tabs must be glued. However, these patterns can be used to assemble a tetrahedron and cube without the need to use tape or glue. The technique used is known as *plaiting*. With proper instructions each student should be able to assemble his own models.

<small>CONSTRUCTION</small>

Carefully cut the patterns out of paper and crease accurately on each line marked. Begin with the labeled faces up and assemble by folding down such that only the numbered faces appear on the surface of the completed model.

<small>REGULAR TETRAHEDRON</small>

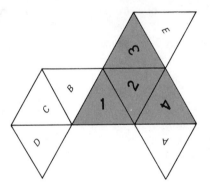

First fold face *A* of the pattern under face 1. Next tuck faces *B*, *C*, and *D* inside faces 1, 2, and 4 such that *C* ends up behind 4 and *D* behind 1 with *B* forming the fourth face. Now fold face 3 over face *B* and tuck face *E* in the slot under face 4 to complete the model.

CUBE

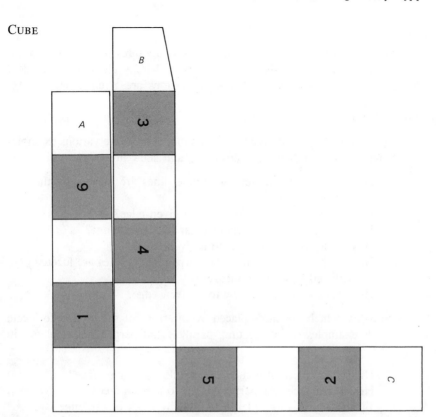

Fold strip *A* over strip *B* and under strip *C*, positioning faces 1 and 5 of the cube. Tuck strip *B* over strip *C* and under strip *A*, positioning faces 4 and 6 of the cube. Fold strip *C* over flap *A* and under strip *B* positioning faces 2 and 3. Now tuck flap *B* into the slot under face 1. The cube with faces numbered 1 through 6 should be assembled.

USES

Any discussion or discovery activity that requires a cube or a tetrahedron might begin with these constructions, thereby giving every student his own model to work with. A few suggestions are given here.

Lines and Planes in Space:

1. Use two pencils placed on edges of the model to locate various examples of parallel, perpendicular, intersecting, and skew lines.

> How many different sets of parallel lines are located by the edges of a cube?
> How many edges are parallel to a given edge?
> How many edges are perpendicular to a given edge?
> How many edges are skew to a given edge?
> How many sets of parallel and perpendicular lines are located by the edges of a tetrahedron?
> How many edges are skew to a given edge?

2. Use 3- by 5-inch file cards placed on the faces of the models to locate various examples of intersecting, parallel, and perpendicular planes in space.

> How many faces of a cube are parallel to a given face?
> How many faces are perpendicular to a given face?
> Illustrate how three planes located by faces of the cube can have a point as their intersection.
> Illustrate three planes that have the empty set as their intersection.
> How many faces are parallel or perpendicular to a given face of a tetrahedron?

Vertices, Faces, and Edges:

3. Label the vertices of each solid and name their various faces and edges.
4. Count the number of vertices, faces, and edges on each solid and relate the numbers to Euler's formula.

Volumes and Surface Areas:

5. Measure with a ruler to find the area of a face of each solid.
6. Find the surface area and volume of each solid.

SUGGESTED EXTENSIONS

1. Folding the other regular polyhedrons is more difficult. This pattern or net of triangles can be folded into an octahedron with its eight faces numbered. Start by folding faces *A* and *B* under faces 1 and 2.

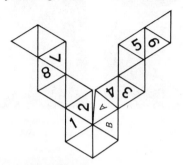

2. Here is a puzzle for students who enjoy this type of activity. Show how an 8-inch strip of paper 1 inch wide can be folded to form a cube. For a somewhat harder version, mark eight ✕'s on one side of the strip and fold a cube with all faces showing ✕'s.

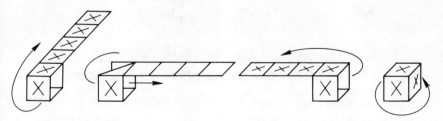

Folding a model of a truncated tetrahedron

1. Start with a 12-inch circle cut from a sheet of newspaper.

2. Fold two diameters to locate the center of the circle.

3. Take the edge of the circle and fold it back onto the center three times in such a way as to form an equilateral triangle.

4. Fold one vertex to the midpoint of the opposite side for an isosceles trapezoid. Fold over two vertices for a rhombus and all three for a small equilateral triangle.

5. Open the model up so that the four equilateral triangles form the faces of a regular tetrahedron.

6. When you fold each vertex of the large triangle into the center, a regular hexagon is formed.

7. Now rest the folded paper loosely in the palm of your hand.

8. Tuck the three upper flaps together to form a model of a truncated tetrahedron.

As the steps in this construction are demonstrated by the teacher, each student can follow them in assembling his own model. Stop after each step to discuss the mathematical properties involved at that stage. Some of the many related questions and answers illustrated by this construction are given here.

1. How can you find the center of a circle?   (step 2)

> By the intersection point of two folded diameters.

2. How can you form a perpendicular bisector of a radius?   (step 3)

> By folding the endpoint of the radius back onto the center point.

3. What is the area of an equilateral triangle?   (step 3)

> One half the base times the height.
> Approximate this area by measuring the model.

4. How does the common intersection point divide the altitudes of an equilateral triangle?   (step 3)

> In the ratio of 1 to 2 to the side and to the opposite vertex.

5. What are the properties of an isosceles trapezoid, a rhombus, and an equilateral triangle?   (step 4)

> Nonparallel sides congruent; base angles congruent.
> All four sides congruent; opposite angles congruent.
> All three sides and all three angles congruent.

6. What are the areas of the isosceles trapezoid, rhombus, and small equilateral triangles in relation to that of the original triangle?   (step 4)

> Three-fourths, one-half, and one-fourth the area of the original triangle.

7. What is the volume of a regular tetrahedron? (step 5)

    One-half the area of the base times the height.

8. What is the size of each interior angle of a regular hexagon? (step 6)

    $120°$.

9. How does the area of the hexagon compare with that of the original triangle? (step 6)

    Two-thirds the original area.

10. How does the surface area of this truncated tetrahedron compare with that of the regular tetrahedron in step 5? (step 8)

    Seven-ninths that of the original surface area. This can easily be seen by studying areas of the nine small triangles formed in step 6.

11. What is the volume of this truncated tetrahedron? (step 8)

    The height of the small tetrahedron removed from the top is two-thirds that of the original tetrahedron. Since they are similar in shape, their volumes must be in the ratio of $(\frac{2}{3})^3$, or $\frac{8}{27}$. The difference, which is the volume of the truncated tetrahedron, must therefore be $\frac{19}{27}$ of the original volume.

    Approximate the volume by measuring the model.

Truncation

To truncate a regular solid means to cut off corners. Truncating the regular polyhedrons can produce some of the most impressive and beautiful of the semiregular solids. In general, these have faces of more than one shape but identical corners.

The value of such models in the mathematics class is inherent in their special shapes and properties. But indeed, even of more value is the skill in space perception that comes from imagining certain of these solids formed from truncating others. This activity is suggested with that in mind. Designed for the better student, it can serve as an exciting, challenging exercise in visualization which culminates, rather than begins, with the model.

DISCUSSION

Initial work begins with a discussion of a cube. Develop the discussion as far as possible without the aid of any models. Encourage students to tax their visualization powers to the fullest. Use a model of a cube only as final verification of initial conjectures.

1. How many faces are on a cube?
2. How many vertices are on a cube?
3. How many edges are on a cube?
4. Imagine each edge marked at its midpoint and each corner cut off with a plane passing through the midpoints of the three corresponding edges.
5. How many faces does the resulting solid have once truncated this way?
6. Are all the faces of the new solid the same shape? If not, describe their various shapes.
7. How many vertices does the new solid have?
8. Describe the shapes of the faces at each corner.
9. How many edges does the new solid have?
10. Describe the faces that come together at each edge.

CONSTRUCTION

The solid formed is a *cuboctahedron* that consists of 6 square faces resulting from the original 6 faces of the cube and 8 triangular faces resulting from the 8 corners cut off. In all there are 12 vertices, 14 faces, and 24 edges.

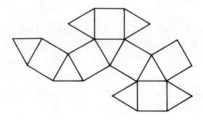

SUGGESTED EXTENSIONS

If you cut off the corners of the cube in stages, starting first with very small pieces, an interesting sequence of solids is formed.

Stage 1:	cube	6 square faces
Stage 2:	truncated cube	6 octagonal and 8 triangular faces
Stage 3:	cuboctahedron	6 square and 8 triangular faces
Stage 4:	truncated octahedron	6 square and 8 hexagonal faces
Stage 5:	octahedron	8 triangular faces

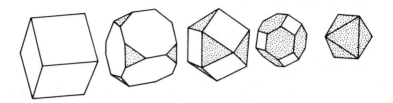

The original 6 faces become octagonal, then square again, and finally disappear. The original 8 corners become triangular, then hexagonal, and finally triangular again. Surprisingly, if you start with the regular octahedron and begin truncating corners, the reverse sequence is formed, and the final solid is again a cube. This describes why the solid in stage 3 is given the combined name cuboctahedron.

The construction of a complete set of these models would make an interesting special project.

5.4 Aids for algebra

Algebraic concepts, by their very nature, are abstract. Any aids that can make them more concrete deserve attention. The illustrations given here suggest ways that flow charts can be used in emphasizing the role of parentheses, to offer an informal method of solving equations, and to introduce sequences.

The geometric approach suggested for factoring trinomials not only presents the topic in a very special visual, manipulative form but has historical significance as well. The ancient Greeks solved all their algebraic problems by geometric means; algebra as we know it had not yet been invented. An interesting research problem is to study how the Greeks solved algebraic equations by geometric means.

Flow charts to teach the use of parentheses

Parentheses in an arithmetic or algebraic expression help determine the order or sequence of operations to follow. Their importance is emphasized in this activity with the aid of flow charts.

CONSTRUCTION

Cut out a series of rectangular and circular pieces of paper and label them with operation commands and inputs. The specific samples illustrated make use of these commands and inputs. But they can be easily modified and expanded to include other inputs, such as decimals, fractions, and integers, and other operations, such as taking square roots. Selected pieces are attached to the bulletin board or chalkboard to form various flow charts.

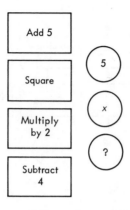

USES

1. Students evaluate the output of each flow chart and express the sequence of operations using parentheses as needed.

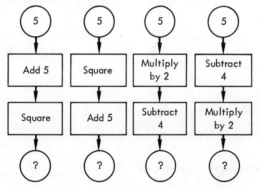

By reordering the operations, expressions are formed with different values, thereby stressing the importance of the use of parentheses.

2. Arithmetic expressions are written on the board. Students then construct the corresponding flow charts from the various inputs and operation commands available. Here are some examples.

$$2 \cdot 5^2 \quad \text{and} \quad (2 \cdot 5)^2$$
$$2 \cdot 5 + 5 \quad \text{and} \quad 2(5 + 5)$$
$$5^2 - 4 \quad \text{and} \quad (5 - 4)^2$$

Valuable experience in the use of parentheses and reinforcement of computational skills come from this simple activity.

3. Introducing a variable into the flow chart sets the stage for developing

algebraic skills with parentheses. Repeat the methods suggested in steps 1 and 2, but use variables this time.

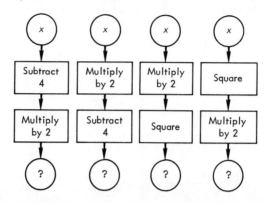

SUGGESTED EXTENSIONS

Place these operations and input on the board. Ask the students to arrange the pieces in as many different flow charts as possible, writing the correct arithmetic expression for each one. There are six possible orderings of the operation pieces; hence six resulting expressions can be formed and evaluated.

| Multiply by 2 | Subtract 4 | Square | 5 | ? |

$$(2 \cdot 5 - 4)^2 = ? \qquad 2(5 - 4)^2 = ? \qquad (2 \cdot 5)^2 - 4 = ?$$
$$[2(5 - 4)]^2 = ? \qquad 2(5^2 - 4) = ? \qquad 2 \cdot 5^2 - 4 = ?$$

With an additional operation (add 5) a total of 24 expressions is possible.

Solving equations with flow charts

Many students have difficulty solving an equation in algebra because they do not know what the equation says nor do they know how to find the particular sequence of steps needed to solve it. A simple, yet surprisingly effective technique for introducing this topic makes use of flow charts. The obvious step-by-step sequencing built into a flow chart is used to analyze the correct sequence needed to solve the corresponding equation.

CONSTRUCTION

Cut out a series of rectangular and circular pieces of paper for inputs and operations. Those shown here will be used in the illustrations that follow, but they can obviously be varied and expanded. However, for each operation given, the corresponding inverse operation is needed.

Selected pieces are attached to the bulletin board or chalkboard to form various flow charts.

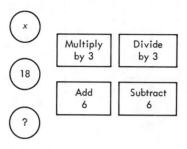

USES

1. Form some flow charts using the available pieces. Students then write the corresponding equations.

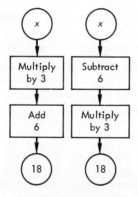

Also write some equations such as these and have the students form the corresponding flow charts.

$$3(x + 6) = 18 \qquad \frac{x}{3} - 6 = 18$$

This activity gives the students a physical format in which to associate abstract equations. Substantial drill at this level can produce significantly better performance in the writing and understanding of equations and word problems.

2. Once the students have this experience in putting equations together, ask how they can be taken apart. It is this basic idea, easily shown with flow charts, that leads to a concrete model for solving equations.

Start with a flow chart and the corresponding equation. Then, beginning with the output, reverse the sequence, performing the inverse operations as you go.

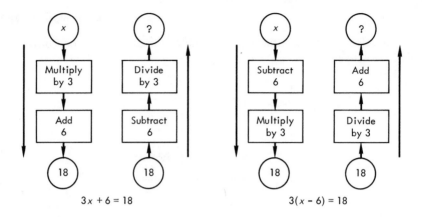

$$3x + 6 = 18 \qquad\qquad 3(x - 6) = 18$$

POSSIBLE EXTENSIONS

1. This method can be extended to the solving of more involved equations. However, it is best suited for informal introduction to the topic.
2. If preferred, flow-chart pieces can be cut from acetate and arranged and projected from the overhead projector.
3. A further modification involves cutting the two flow-chart patterns from heavy paper and projecting them on the chalkboard with the overhead projector. As new problems are studied, old inputs and operations are erased from the board. The projected outline of the flow chart remains.

Using flow charts to introduce sequences

All too frequently the topic of sequences and series appears in the algebra program with little, if any, prior exposure. So many interesting sequences can be explored informally at an earlier level that they should not be neglected. However, without some concrete method of presentation, many teachers believe that the concept is too abstract. The aid suggested here will be described specifically in introducing the Fibonacci sequence. It can easily be modified for other sequences.

CONSTRUCTION

Cut out circular, rectangular, and diamond-shaped pieces of paper. They are marked in various ways to develop different sequences. The Fibonacci sequence requires the specific pieces shown.

188

Uses

Set up this flow chart with the pieces shown. They can be attached to the bulletin board or chalkboard or, if preferred, drawn on acetate and projected on the overhead projector.

1. Discuss the meaning of the diamond-shaped decision box. It asks the question and directs the sequence of steps one way for the answer *yes* and another way for the answer *no*. Note also the possible loop built into the program. As long as the answer to the question is no, the sequence of steps will repeatedly loop around. Only when the answer is yes will it break out of the loop and stop.

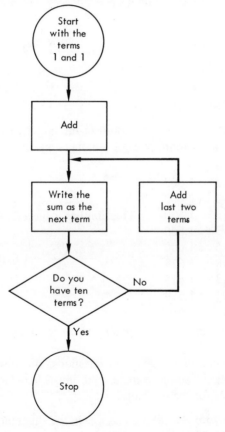

2. Have the students write the sequence formed, adding a new term each time the program loops back. The first 10 terms in the Fibonacci sequence are the result:

$$1, 1, 2, 3, 5, 8, 13, 21, 34, 55$$

3. Modify the flow chart by replacing the decision box with one of these. The same sequence is produced, but now the last term is determined in a different way.

4. Modify the sequence produced by starting with different terms. These still produce examples of Fibonacci sequences, but not the familiar one.

1, 3, 4, 7, 11, 18, 29, 47, 76, 123   2, 5, 7, 12, 19, 31, 50, 81, 131, 212

SUGGESTED EXTENSIONS

1. Have the students compute the ratio of the larger to smaller number in each successive adjacent pair of numbers in the familiar Fibonacci sequence.

$$
\begin{array}{l}
1 \text{ to } \ 1 = 1.000 \\
2 \text{ to } \ 1 = 2.000 \\
3 \text{ to } \ 2 = 1.500 \\
5 \text{ to } \ 3 = 1.666 \\
8 \text{ to } \ 5 = 1.600 \\
13 \text{ to } \ 8 = 1.625 \\
21 \text{ to } 13 = 1.615 \\
34 \text{ to } 21 = 1.619 \\
55 \text{ to } 34 = 1.617
\end{array}
$$

The limiting value of this ratio is

$$\frac{1 + \sqrt{5}}{2} = 1.618034 \ \ldots$$

2. Discuss the various places in nature where examples of this sequence are found.

3. With other inputs, operations, and questions, students can study other familiar sequences. Some examples are given here, together with suggested questions for the decision box.

7, 14, 21, 28, 35, 42, 49, . . .   Is the difference of the last two
1, 2, 4, 8, 16, 32, 64, . . .   terms greater than 4000?

Is the sum of the terms greater than 500?

Is the ratio of the last two terms greater than .9?

A model for factoring trinomials

The concept that underlies trinomial factoring can be demonstrated and the skills reinforced with this teaching aid, which offers a concrete, visual, geometric model for an abstract algebraic idea.

CONSTRUCTION

Cut out a series of 5 x 5 squares, 1 x 5 rectangles, and 1 x 1 squares. Label the longer dimensions x and the shorter ones 1.

USES

The area of the larger squares is called x^2.
The area of the rectangles is called x.
The area of the smaller squares is 1.

Monomials, binomials, and trinomials in x can now be represented by the appropriate geometric figures attached to the bulletin board or chalkboard.

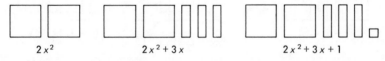

$2x^2$ $\qquad$ $2x^2 + 3x$ $\qquad$ $2x^2 + 3x + 1$

In factoring a trinomial, the various monomial parts are arranged in a rectangular shape. The dimensions of the rectangle give the factors.

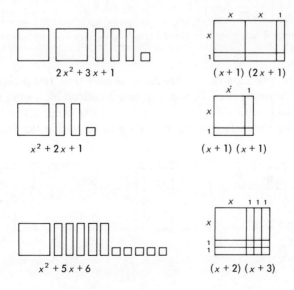

$2x^2 + 3x + 1$ $\qquad$ $(x+1)(2x+1)$

$x^2 + 2x + 1$ $\qquad$ $(x+1)(x+1)$

$x^2 + 5x + 6$ $\qquad$ $(x+2)(x+3)$

Small colored acetate figures on an overhead projector can be used instead of paper figures on the board. Students can also try drawing the appropriate figures on graph paper at their seats.

Obvious restrictions are implied in this aid. Positive integral values are needed and only reasonably small numerical coefficients are practical. Trinomials that cannot be factored into integral factors produce pieces that cannot be arranged into a rectangular display.

Suggested Extension

Have areas that represent positive coefficients be cut from white paper and those that represent negative coefficients be cut from red paper. Now the factoring of trinomials with both positive and negative coefficients can be illustrated. Negative values add out with positive values, so red areas are placed on top of white areas. Equal positive and negative areas can be added where needed.

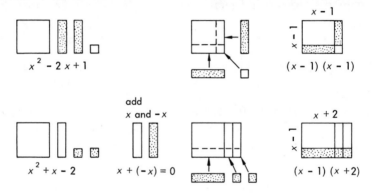

While the multiplication of two binomials usually does not present the difficulties that factoring does, it, too, can be introduced by a similar method. Draw a rectangle with dimensions those of the two binomial factors given. Then subdivide the rectangle into x^2's, x's, and 1's to find the product.

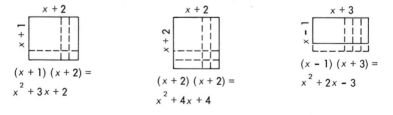

Exercises

1. Show how the numbers 0 through 10 can be expressed from the four digits in this year, using each digit only once together with the four fundamental operations, raising to a power, and/or extracting a square root.

2. Sometimes dates are marked on labels of containers by notching in various positions. Prepare a key to show how this technique can be used to give a date of month, day, and year. Then show how to notch a label to record your birthdate.

3. An unlimited number of squares and rectangles measuring 2 x 2, 2 x 4, 2 x 6, 4 x 4, 4 x 6, and 6 x 6 are available. How many different-sized rectangular prisms can be constructed from them?

4. A cone is formed from a circular piece of paper. When rolled up in different ways, different cones are formed but all have the same slant height. Draw a graph to show the relationship between each of the following:
   (a) the radius and the height
   (b) the radius and the volume
   (c) the radius and the surface area

5. Prove that only five regular polyhedrons are possible.

6. Find the volume and the surface area of the truncated tetrahedron constructed from a 12-inch circular piece of paper as described on page 180.

7. In the same construction, prove, using the conventional techniques of geometric proof, that the figure formed in step 3 is an equilateral triangle and in step 6 is a regular hexagon.

8. Find the volume and surface area of the cuboctahedron formed from a 4-inch cube.

9. Discuss the duality property as it can be applied to the cube and the octahedron.

10. Write all the different algebraic expressions that can be formed from an input of 3 and the operations "add 2," "cube," and "multiply by $\frac{1}{2}$" in a flow chart.

Activities

1. Read and report on Chapter 8, "Using Models and Instructional Aids," by Donovan A. Johnson, Emil J. Berger, and Gerald R. Rising in the Thirty-fourth Yearbook of the National Council of Teachers of Mathematics, *Instructional Aids in Mathematics* (Washington, D.C.: N.C.T.M., 1970).

2. Describe several different types of mathematical games or activities to help develop arithmetic skills that might be played in class with a deck of cards labeled 1 through 10.

3. Compare the method of multiplication using Napier's rods to the gelosia method described in the Treviso arithmetic (1478).

4. Prepare the per cent chart described on page 158 on a transparency and discuss how it might be used in teaching a lesson on per cents.

5. Prepare a set of 16 cards punched so that they can be sorted in just four operations. Then describe how, once shuffled, they can be used to spell out the message MATHEMATICAL FUN.

6. Use a string knotted at 1-foot intervals to lay out a 10-foot square as the ancient Egyptians might have done.

7. Read and report on President Garfield's proof of the Pythagorean theorem.

8. Construct the tetratetraflexagon described on page 168.

9. Construct a hexahexaflexagon.

10. Plan a field activity for a junior high class that would make use of a hypsometer.

11. Prepare a report on some of the history surrounding the regular polyhedrons and famous mathematicians, such as Pythagoras, Plato, Euclid, Archimedes, and Kepler, who studied them.

12. Construct a cube by the plaiting technique described on page 177.

13. Make a set of models showing various stages in the truncating of a cube to an octahedron as shown on page 183. Start with a 4-inch cube.

14. Describe five different junior high classroom activities that could center around the student's construction or use of a cube at his desk. Make some metric in nature and others nonmetric.

15. Discuss how the construction shown on page 180 could be used in a junior high class, an algebra class, and a geometry class.

16. Describe a lesson that would make use of flow charts in developing the role of parentheses in algebraic expressions.

17. Prepare a report on the apparent relationship between the Fibonacci sequence and certain aspects of nature.

18. Plan a bulletin board display on the "golden ratio."

19. Discuss the advantages in having an aid to demonstrate the algebraic concept of factoring.

20. Construct a teaching aid other than those described in this chapter that can be used in teaching algebra.

References and selected readings

BRUYN, D. L. *Geometrical Models.* Portland, Me.: J. Weston Walch, 1963.

CUNDY, H. MARTYN, and A. P. ROLLETT. *Mathematical Models.* New York: Oxford University Press, Inc., 1961.

HOLDEN, ALAN. *Shapes, Space, and Symmetry.* New York: Columbia University Press, 1971.

KRULIK, STEPHEN. *A Handbook of Aids for Teaching Junior–Senior High School Mathematics.* Philadelphia: W. B. Saunders Company, 1970.

———. *A Mathematics Laboratory Handbook for Secondary School.* Philadelphia: W. B. Saunders Company, 1972.

Mathematics Teacher. "Collapsible Models of the Regular Octahedron" (Oct. 1972), p. 530.

———. "The Five Platonic Solids" (Jan. 1969), p. 42.

———. "A Look at Regular and Semiregular Polyhedra" (Dec. 1972), p. 713.

———. "A Physical Model for Factoring Quadratic Polynomials" (Mar. 1972), p. 201.

———. "Tessellations" (Apr. 1973), p. 339.

———. "A Tiny Treasury of Tessellations" (Feb. 1968), p. 114.

National Council of Teachers of Mathematics. *Instructional Aids in Mathematics* (Thirty-fourth Yearbook). Washington, D.C.: N.C.T.M., 1970, Chap. 8, "Using Models as Instructional Aids," by Donovan A. Johnson, Emil J. Berger, and Gerald R. Rising.

CHAPTER SIX

AUDIO-VISUAL FACILITIES, RESOURCES, AND REFERENCES

One aspect of aids, activities, and ideas for the teacher in the mathematics classroom centers around familiar audiovisual facilities. These include use of films, filmstrips, film loops, slides, and video tapes as well as equipment such as the opaque and overhead projectors. Films can serve many different roles in the classroom. They can introduce new topics and illustrate equipment and applications not available within the classroom. They can motivate the study of mathematics and can serve as a change of pace in the class routine. An extensive listing of audiovisual materials available from many sources can be found in this publication of the N.C.T.M.:

RAAB, JOSEPH A. *Audiovisual Materials in Mathematics.* Washington, D.C.: National Council of Teachers of Mathematics, 1971.

This pamphlet contains a listing of approximately 5000 films, filmstrips, film loops, video tapes, and transparencies available from approximately 200 producers and distributors. In addition, new audiovisual materials are regularly reviewed in the *Mathematics Teacher.*

A wide variety of other types of materials, models, and aids for use in the mathematics classroom is available commercially. Some selected sources of these supplies are listed here.

Creative Publications, P.O. Box 10328, Palo Alto, Calif. 94301. Supplies a wide assortment of mathematics curriculum materials of a novel nature especially suited for laboratory use.

J. L. Hammett Company, Hammett Place, Braintree, Mass. 02184. Distributes most of the common supplies and materials needed in a classroom laboratory.

LaPine Scientific Company, 6001 S. Knox Ave., Chicago, Ill. 60629. An excellent source of teaching aids for the mathematics classroom.

Math-Master, Box 310, Big Spring, Tex. 79721. Source of filmstrips, transparencies, and other mathematical models and aids.

Sargent-Welch Scientific Company, 7300 N. Linder Ave., Skokie, Ill. 60076. Extensive supplies of models and visual aids.

J. Weston Walch, Box 1075, Main Post Office, Portland, Me. 04104. Supplier of a wide variety of manuals, posters, and workbooks for both student and teacher.

The remainder of this chapter is devoted to making the most effective use of typical classroom facilities. Simple techniques for making transparencies for the overhead projector are described along with suggestions and ideas for using the bulletin board and chalkboard. Finally, the question of the teacher's adapting his own individual classroom so that it can be made more suitable for an occasional math-lab approach to teaching is considered. Although most teachers have some supplies available in their classes, suggestions are given for building up a collection of the materials most useful for laboratory-type activities. These range from basic supplies to simple models to mathematical games to suggested library references for the students.

A valuable reference on the use of audiovisual facilities, resources, and references can be found in the Thirty-fourth Yearbook of the N.C.T.M.:

National Council of Teachers of Mathematics. *Instructional Aids in Mathematics* (Thirty-fourth Yearbook). Washington, D.C.: N.C.T.M., 1973.

6.1 Uses of the overhead projector

Of all the audiovisual equipment available today, none has greater potential in the mathematics classroom than the overhead projector. Its simplicity, versatility, and adaptability make it an indispensible aid for the teacher. Unlike so many types of visual projectors, a variety of dynamic, imaginative, and original materials can be prepared for the overhead projector by the teacher easily and readily with a minimum of tools and talent. This section describes the supplies needed and some of the mechanical techniques that can be utilized in preparing transparencies to use with an overhead projector in the mathematics classroom.

These materials are essential for transparency construction:

clear acetate sheets—preferably 8½ by 11 inches, large enough
to cover most of the projecting surface
acetate marking pens—in assorted colors with ink designed ex-
pressly for adhering to acetate

Additional supplies can prove helpful in constructing neat, accurate, and
creative transparencies:

ruler, protractor, 30-60-90 triangle—made from clear plastic so
that they themselves can be projected
colored acetate—available in sheets and useful to focus attention
and contrast different regions
heavy paper or tagboard—for preparing masks and cutouts
razor blades or knife—for making special cuts in the acetate
compass—must be adaptable to hold marking pens
masking tape—for hinging overlays to basic transparency
graph paper—to use under the acetate for proper spacing of figures and neat, straight lettering

With these supplies and a bit of imagination, many ideas can be developed in the format of transparencies. Some can lead to new discoveries,
others can help reinforce old ones. Some can emphasize a sequence of
steps in a single development; others can offer a variety of examples of a
single concept. Some can be designed to review arithmetic skills; others
strengthen geometric concepts or develop space perception.

There is a wide variety of mechanical techniques that can be used in
the preparation of transparencies. Each of the examples that follows utilizes
a different technique in meeting its objective.

Flow charts for the overhead projector

Flow charts offer to transparencies a new format for the review of
arithmetic skills. The flow-chart format is easily adaptable to addition,
subtraction, multiplication, and division using whole numbers, integers, or
rational numbers expressed as fractions, decimals, or per cents.

Step 1: Start with a sheet of heavy paper on tagboard. Cut out two input
and output circles and two instruction boxes each about 1¼ inches
high. These, along with the small triangles that indicate arrows,
can best be cut out with a razor blade.

Step 2: Now cut four strips of clear acetate about 1½ inches wide. Then
cut two slots in the original mask on each side of each opening,

as shown. Make them just long enough to allow the strips to slide easily inside them.

Step 3: Slide the strips through the slots such that they are on top of the paper on both sides and under where the openings have been cut.

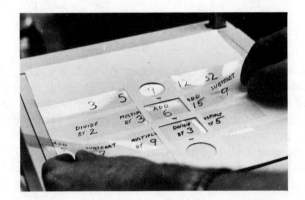

Step 4: Label the input and operation strips according to the particular use planned for the transparency. Several suggestions are given.

WHOLE-NUMBER REVIEW

Input	0	1	4	12	36
Operation 1	Multiply by 2	Add 6	Subtract 3	Divide by 2	Square
Operation 2	Add 8	Subtract 9	Divide by 3	Square	Multiply by 3

FRACTION REVIEW

Input	0	$\frac{1}{2}$	$\frac{3}{4}$	$\frac{1}{3}$	$1\frac{1}{8}$
Operation 1	Add $\frac{1}{2}$	Subtract $\frac{1}{8}$	Add $\frac{3}{4}$	Multiply by 2	Divide by 2
Operation 2	Multiply by $\frac{1}{3}$	Divide by $\frac{1}{2}$	Multiply by 3	Add $1\frac{1}{3}$	Subtract $\frac{1}{4}$

INTEGER REVIEW

Input	−1	+3	−6	+10	−15
Operation 1	Multiply by −2	Add −3	Subtract +6	Divide by +3	Square
Operation 2	Add +5	Square	Divide by −3	Subtract −2	Multiply by −4

Any pair of operations can be matched with any input. With three strips containing five entries each, a total of 125 problems can be formed simply and quickly. This makes the flow-chart format useful for the practice of rapid mental computation and for classroom competition. A more challenging but very effective modification is to remove the input strip and use it as the output strip. Students then have to start with the output or answer and work back through using inverse operations to find the original input.

Axes translation using transparencies

Concepts in graphing can be dramatically and vividly illustrated with the overhead projector by using the graph of a function and the corresponding set of axes on different sheets of acetate. Here is an illustration in which absolute values are used. Begin by constructing two acetate sheets.

Sheet 1: A conventional set of axes in one color.
Sheet 2: The graph of the function $y = |x|$ in another color.

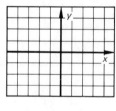

Sheet 1   Sheet 2

When sheet 2 is placed on top of sheet 1, the basic graph is projected.

$y = |x|$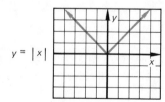

When sheet 2 is flipped about the horizontal axis, a new graph is projected.

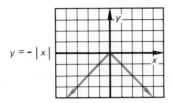

$$y = -|x|$$

Rotating 90° clockwise or counterclockwise illustrates two other graphs:

$$|y| = x \quad \text{and} \quad |y| = -x$$

Displacing sheet 2 one unit up from its original position illustrates another graph.

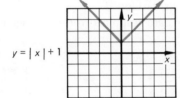

$$y = |x| + 1$$

Displacing sheet 2 two units down from the original describes still another graph.

$$y = |x| - 2$$

At this point other positions of vertical translation should be illustrated, with the class giving the proper equations. This first concept can then be further reinforced by offering the equation and having the student position the graph properly. The effect of vertical translation and the constant a in $y = |x| + a$ is thus illustrated.

Sheet 2 is now moved one unit to the right of its original position and the corresponding equation developed.

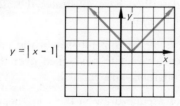

$$y = |x - 1|$$

Shifting sheet 2 two units to the left of the original position yields another graph.

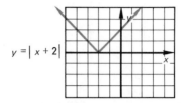

$y = |x + 2|$

Again the horizontal translation is reviewed with additional examples by giving new equations and by giving new positions. The role of the constant b in $y = |x + b|$ is thus developed.

Now the two translations are combined for still other graphs.

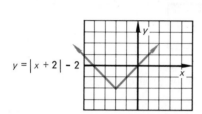

$y = |x + 2| - 2$

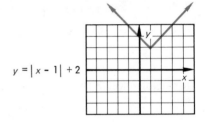

$y = |x - 1| + 2$

These concepts can be reviewed with more positions and more equations until the student grasps the concepts involved in graphing equations of the form $y = |x + b| + a$.

A nice variation of this development can follow for reinforcement of concepts by placing sheet 1 on top of sheet 2 and moving the axes rather than the graph. The whole development is, of course, readily adaptable to other functions, such as these:

$y = 2x$ straight line $y = x^2$ parabola

Sketching prisms through a transparency sequence

The transparency set described here illustrates the steps in sketching prisms using a sequence of overlays.

Step 1: Draw oblique views of the upper bases of the prisms.

Step 2: On a separate sheet of acetate positioned over the first, draw in the lateral edges congruent and parallel.

Step 3: On a third sheet of acetate draw in the congruent bases.

Step 4: Use masking tape to hinge the three sheets of acetate together at the top.

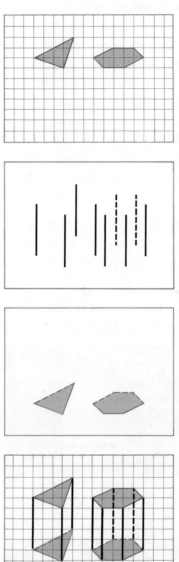

Students first watch the complete sequence and then follow it a second time, copying each step at their seats. The transparencies can then be used to emphasize that the upper and lower bases are congruent, as are the lateral edges. By following these steps, students can then sketch other prisms.

Pattern discovery using cutout masks

In this example a cutout in a mask is used to isolate a selected square array of numbers from a given set. Students try to discover how to find the sum of the numbers in each square array without adding them all. Construction of the transparency aid is very simple.

Step 1: The calendar for the month is copied on a sheet of acetate. Since accurate spacing is needed, use a piece of graph paper under the acetate for horizontal and vertical alignment of the numbers.

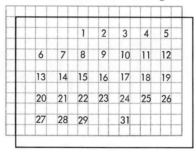

Step 2: With a razor blade a square is cut from a sheet of heavy paper or tagboard large enough to just expose a 3 x 3 array of numbers.

Step 3: By placing this mask over the number grid of the calendar, only a 3 x 3 array will be projected. Repositioning the mask simply exposes a new array.

Students are asked to add the numbers in various arrays and then try to discover the relationship between each sum and the particular numbers exposed. Various suggestions should be encouraged and tested. The discovery hopefully will come from the students, if necessary with helpful hints. The sum of the numbers in each 3 x 3 array will always be 9 times the center number.

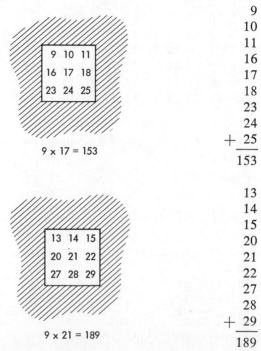

9 x 17 = 153

9
10
11
16
17
18
23
24
+ 25
153

9 x 21 = 189

13
14
15
20
21
22
27
28
+ 29
189

Encourage the students to verify the discovery algebraically. If the middle number is a, then the nine numbers can always be expressed this way regardless of where the mask is placed on the calendar. Illustrate this display with another transparency.

The sum of these numbers is obviously $9a$.

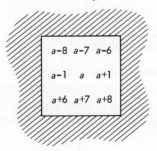

A clever variation of this idea involves the addition table. Cut masks with 2 x 2, 3 x 3, 4 x 4, and 5 x 5 openings. Place these masks over the 10 x 10

```
0  1  2  3  4  5  6  7  8  9
1  2  3  4  5  6  7  8  9 10
2  3  4  5  6  7  8  9 10 11
3  4  5  6  7  8  9 10 11 12
4  5  6  7  8  9 10 11 12 13
5  6  7  8  9 10 11 12 13 14
6  7  8  9 10 11 12 13 14 15
7  8  9 10 11 12 13 14 15 16
8  9 10 11 12 13 14 15 16 17
9 10 11 12 13 14 15 16 17 18
```

addition table and see if the students can discover the relationship between the sum and the size of the array and the repeating number in the diagonal.

5 6		9 10 11
6 7		10 11 12
		11 12 13

sum: $4 \times 6 = 24$ sum: $9 \times 11 = 99$

		8 9 10 11 12
12 13 14 15		9 10 11 12 13
13 14 15 16		10 11 12 13 14
14 15 16 17		11 12 13 14 15
15 16 17 18		12 13 14 15 16

sum: $16 \times 15 = 240$ sum: $25 \times 12 = 300$

Once the discovery is made and verified, the mask can be removed from the number grid so that the entire 10 x 10 array is projected. Students can then compete to see who can find the total sum first.

sum: $100 \times 9 = 900$

The square is 10×10 or 100. The diagonal repeats 9's. Hence their product, 100×9, gives the sum, 900.

Rotating disks for problem variation

Movable disks can easily be constructed on a transparency. Not only do they add an interesting variety and dimension to the potential of the overhead, but they can put a familiar computational problem in a new and stimulating format. One possible use related to a gasoline gauge is illustrated here.

Step 1: The scales are first copied on a clear sheet of acetate.

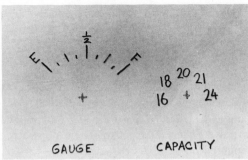

Step 2: Cut two disks of heavy paper, one with a pointer and the other with a hole cut out. The size of the disks is determined by the size of the scales in step 1.

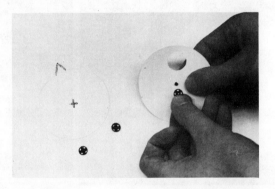

Step 3: Cut a small cross on the disks and the acetate where the pivots should be. Align the disks over the acetate and attach them with small sewing snaps as shown. If the snaps don't fit, cut out additional space from the acetate.

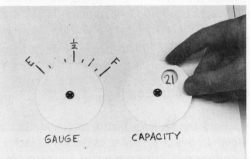

Step 4: The disks should rotate easily. Move them to a desired position and ask related questions such as those on the next page.

How much gas is left in the tank?

$$\tfrac{1}{4} \times 21 = 5\tfrac{1}{4} \qquad 5\tfrac{1}{4} \text{ gallons}$$

At 16 miles per gallon, how far will the car be able to drive?

$$5\tfrac{1}{4} \times 16 = 84 \qquad 84 \text{ miles}$$

By varying the tank capacity and/or the gauge, a variety of computa-tional problems evolve.

In the example just described, the disks were opaque and the background transparent. The reverse technique is illustrated here with transparent disks mounted on an opaque background.

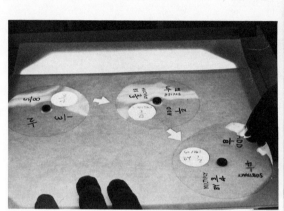

Skills with fractions are reviewed by moving the marked disks in different positions for different inputs and operations. With four choices on each disk, a total of 64 different settings are possible. Again, the acetate disks are attached to the backing mask with sewing snaps.

6.2   Uses of the bulletin board

In many classrooms the bulletin board serves as the focal point for exhibits and posters. In others it serves, as does the blackboard, in the actual teaching and on-the-spot development of an idea. In some it displays daily mathematical tidbits, historical quotes, or puzzle problems. In still others it is the source of enrichment materials related to the unit on hand. Conveniently available in most classrooms, the bulletin board can serve an important role in the overall learning experience. To illustrate the potential of the bulletin board, several illustrations of possible uses are given here. In general, its use is restricted only by the imagination of the teacher using it.

Bulletin board displays

A topic well suited for a bulletin board display in the mathematics classroom is optical illusions. Almost everyone enjoys a good display of these. They can be effectively exhibited just for fun or they can be used to convince students that "seeing is not always believing." This can be an important point to make in geometry where students are encouraged in their proofs to rely on reason rather than relying too heavily on figures.

Some of the many popular illusions are given in this collection. A new one might be added each week and students encouraged to search for more on their own.

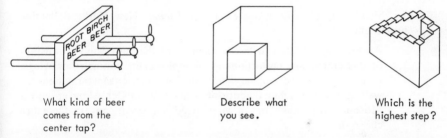

What kind of beer comes from the center tap?

Describe what you see.

Which is the highest step?

First impressions in comparing segments a and b can lead to poor judgments in these illusions.

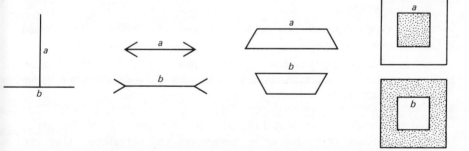

Of course, when the optical illusions are taken down, they should be stored in a box such as the one shown on the next page.

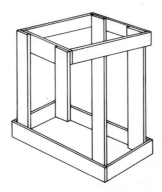

Some additional sources of optical illusions and related topics are listed here. Each represents a decidedly different approach to illusions.

BRANDES, LOUIS G. *An Introduction to Optical Illusions.* Portland, Me.: J. Weston Walch, 1957.

> A vivid collection of some of the more common optical illusions.

ESCHER, M. C. *The Graphic Work of M. C. Escher.* New York: Ballantine Books, Inc., 1967.

> An exciting collection with commentary of the most famous drawings of this craftsman. Distortions of natural scenes, symmetry drawings, regular divisions of a plane, and unique geometric transformations, as well as vivid sketches of curiosities such as ants crawling around a Moebius strip make this a guaranteed success in any mathematics classroom.

HUFF, DARRELL, and IRVING GEIS. *How To Lie with Statistics.* New York: W. W. Norton & Company, Inc., 1964.

> Contains some interesting illustrations of illusions or deceptions in advertising.

OSTER, GERALD. *The Science of Moiré Patterns.* Barrington, N.J.: Edmund Scientific Co., 1965.

> An interesting discussion of Moiré patterns available with a wide assortment of posters, slides, and kits. Certain to be exciting in a mathematics class.

Students should be encouraged to search for their own optical illusions as well since these often appear in newspapers and periodicals. They can prepare bulletin board displays themselves, including some of the deceptions in advertising through use of misleading graphs and figures.

Posters

Several companies have available collections of posters suitable for display in the mathematics class either as enrichment material, puzzle posers, or artistic decorations.

Creative Publications, P.O. Box 10328, Palo Alto, Calif. 94301
Excellent collections of vividly colored posters with puzzles and designs certain to catch the eye of every student.

J. Weston Walch, Box 1075, Main Post Office, Portland, Me. 04104
A wide assortment of poster sets covering many different aspects, such as applications and history of mathematics.

Individual posters made by the teacher or student can bring life to the bulletin board as well. They should be eye catching, with vivid colors and snappy titles, and simply stated questions.

REFLECTIONS

READ THIS MESSAGE SEEN IN A MIRROR. CAN YOU FIND THE EIGTH MISTAKES IN IT?

DART TIME

5
10
15
20
25

FIND ALL THE SCORES POSSIBLE WITH JUST **4** DARTS

This clever poster poses an interesting question as well as showing an interesting mistake. Once posted it can serve to motivate a class activity.

Source: Max A. Sobel, Evan M. Maletsky, and T. Hill, *Essentials of Mathematics,* Book 2 (Lexington, Mass.: Ginn and Company, 1973), p. 391.

Let five students all try shaking hands with each other. Then do the same with just four, three, and two students. This experiment and subsequent discussion can then lead into a nice mathematical discovery.

people	handshakes
2	1
3	3
4	6
5	10
.	.
.	.
.	.
n	$\dfrac{n(n+1)}{2}$

Now that the generalization is made, students should look for the mistake in the poster.

Student projects

Here is a special project on tessellations made by a student and effectively displayed on the bulletin board.

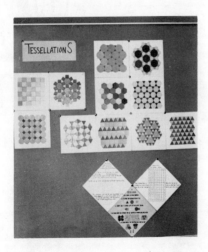

Historical items

In a slightly different vein, interesting bits from the history of mathematics can be presented via the bulletin board. Stories about Thales, Pythagoras, Euclid, and Archimedes make interesting reading when studying geometry. Or pictures of Descartes, Pascal, and Gauss can be posted when studying coordinate geometry, probability, and complex numbers. For students interested in philately, have them search out and display some of the stamps relating to mathematics and mathematicians. It is surprising how exciting this can be to some students.

In the illustrations of stamps shown here, the Scott catalog numbers are given in parentheses for each stamp. All the stamps shown are reasonably inexpensive.

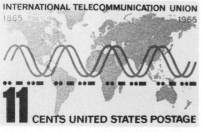

Sine curve (U.S. 1274)   Moebius strip (Brazil 1053)

Pythagorean theorem
(Nicaragua C762)   Formulas (Israel 474)

Riese (Germany 799)   Euler (Switzerland B267)

Stevin (Belgium B321)

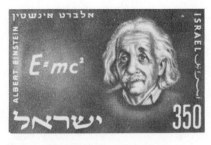

Einstein (Israel 117)

Gauss (Germany 725)

Descartes (France 330)

Pascal
(France B181)

Bolyai
(Rumania 1345)

Chebyshev
(Russia 1050)

As a teaching aid

The bulletin board can also function strictly as an immediate aid in teaching as illustrated with this example on regular tetrahedrons. Congruent equilateral triangles measuring about 2 inches on each edge are prepared. Students then set up the various possible patterns or nets of four triangles by tacking triangles on the bulletin board. Only three possible arrangements exist if rotations and reflections are not counted. And only two of those form patterns for a tetrahedron.

Exploration

Interesting discovery problems can also be illustrated on the bulletin board. One challenging example is the shortest-path problem. Three tacks are placed on the bulletin board in a triangular array and a reasonable distance apart. Equipped with a string and a measuring tape, students try to find the shortest path connecting the points. Intermediate points are allowed.

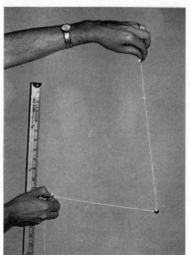

The problem is not as simple as it may sound, but this will make the students only more anxious to try. The solution is somewhat surprising. An intermediate point that forms three 120° angles with the given points locates the shortest path. If the triangle itself contains an angle of 120° or more, the two shortest sides of the triangle form the best path.

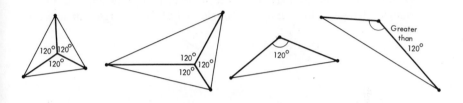

6.3   Uses of the chalkboard

This section is designed to illustrate some of the potential uses of the chalkboard beyond those that are typical and routine. Students are often motivated simply by the new techniques they see used by the teacher in the classroom. With that in mind, these examples are given as suggestions for variation and change of pace as well as for clarification of communication.

Certain materials are basic to good, effective chalkboard use.

> chalk in assorted colors—to attract attention, emphasize key words and numbers, and identify parts of geometric figures
> yardstick, compass, and protractor—for the constuction of accurate figures and to duplicate the standard tools of the student
> coordinate grid—preferably permanent and conveniently located and scored rather than boldly ruled so that chalk-drawn figures are easily seen
> paper or tagboard—for displays attached to the board with masking tape, other adhesives, or magnets

No mathematics class should be without these items, especially if the chalkboard is to be used at its maximum for motivation. Imaginative use of these materials can make a typical lesson unique or a verbal, abstract development visual and concrete. To illustrate this point, the following examples are given.

Describing motion and change

Once the relationships among interior and exterior angles of a triangle have been studied with the usual fixed figures, an angle *A* is drawn on the board. A yardstick is placed on the figure and rotated as if pivoted at *B*. As the yardstick is moved, the vertex *C* of the triangle changes location. The corresponding changes in the various angles are studied with questions such as these.

1. As angle *B* of the triangle decreases in size, how does angle *C* of the triangle change? How small can angle *B* be? How large can angle *C* be?

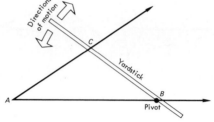

2. As angle *B* of the triangle increases in size, how does angle *C* of the triangle change? How large can angle *B* be? How small can angle *C* be?
3. How many right triangles can be formed? How many isosceles triangles are possible? Under what conditions would triangle *ABC* be equilateral?
4. What is the longest length possible for side *AC*? For side *BC*?
5. How does the exterior angle at *B* change as angle *B* of the triangle decreases? How large can the exterior angle at *B* be?
6. How does the exterior angle at *B* change as angle *C* of the triangle decreases? How small can the exterior angle at *B* be?
7. When are the exterior angles at *B* and *C* the same in size?

The added dimension in this approach is the motion, the change. Students need to watch to see what is happening, and they need to think beyond the confines of a specific diagram of fixed size.

Illustrating loci

Many simple yet interesting loci can be illustrated with the chalkboard. A yardstick can be placed against the side and bottom of the board. Students are first asked to guess at the path of the midpoint *P* as the yardstick is moved but always with its ends on the horizontal and vertical edges. Then the loci, a quarter circle, is traced out by actually moving the yardstick.

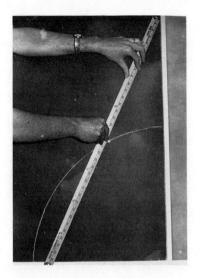

A circular piece of cardboard can be rolled along the chalk tray. Students in turn try to describe the paths of various points punched through the disk.

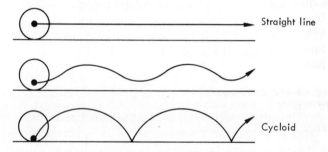

An interesting and challenging extension of this idea involves flipping over other polygons, such as squares and rectangles.

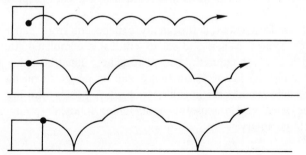

In. each of these cases the path consists of a series of circular arcs with the pivot point the center and with radii that may change from position to position. Bright students can have fun with these loci problems where the square is flipped about a fixed square taped to the chalkboard.

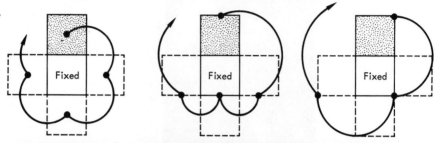

When activities of this type are done with the class, it is most exciting to have the students guess at the paths first before they are actually drawn. Not only can these serve as an interesting introduction or review of loci but as motivating a discussion of periodic functions as well. Once figures are shown, students should draw them accurately at their desks using graph paper and compasses.

Many challenging problems of this type can be given.
> Flip a square about a fixed equilateral triangle.
> Flip an equilateral triangle about a fixed square.
> Flip an equilateral triangle about a fixed equilateral triangle.
> Roll a circle about a fixed circle.

Projecting on the chalkboard

The chalkboard can serve well as a screen for the overhead projector in many types of classroom activities. When complicated geometric diagrams are needed, they can be prepared in advance to save classroom time. Then when projected on the board, various parts can be emphasized with colored chalk. This technique can be especially useful in locating congruent triangles in plane figures and in identifying various parts of space figures.

Layouts or formats used repeatedly can be cut from paper and projected on the board. When one set of entries is completed, the board is erased and a new set entered. The picture here illustrates this technique with a flow chart. Grids of various sorts can be projected on the board as well. These can include rectangular, triangular, polar, semi-log, and log-log. The rectangular grid is useful for sketching solids, graphing equations, and informal geometry.

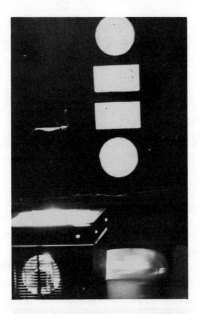

A series of number lines projected on the board can save a great deal of setup time in certain classes. Once the graphs are drawn, simply erase them and begin again; the number lines remain. A slight modification involves mapping from one number line to another to illustrate an operation or function. Project the number lines and get the students involved by having them map the outputs for various inputs. Shown are several examples.

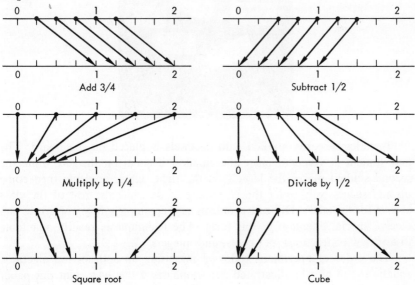

Add 3/4

Subtract 1/2

Multiply by 1/4

Divide by 1/2

Square root

Cube

Reversing the problem can prove interesting, especially for the better students. In this case, the functions are drawn in and the student must identify them.

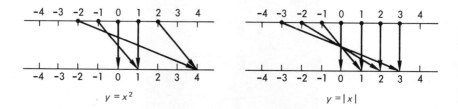

$y = x^2$

$y = |x|$

Displays on the chalkboard

Another potential use of the chalkboard incorporates cards attached to the board with tape, other adhesives, or magnets. This technique can serve as a change of pace, but even more it can dynamically illustrate and emphasize problems involving specific sequences of steps or orderings.

A series of cards marked with decimals is placed on the board. The first student is asked to place the smallest decimal on the left while the second student places the largest on the right. Other students then come up and successively order the remaining cards while the rest of the class checks for mistakes. Difficult problems and common sources of errors can receive special attention in this form. The technique is readily adaptable to a review of fractions, per cents, and integers.

The order of operations followed in setting up and solving algebraic equations can also be illustrated and emphasized this way. Cut out paper with various operations marked on them and show how reordering gives different results. Series can also be introduced in this way. See Chapter 5 for more details.

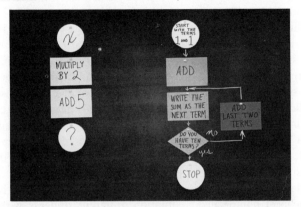

The teacher–actor can prepare something special for class by covering up a key theorem on the board with a sheet of newspaper before class. The dramatic climax comes just at the right moment, when he tears off the paper and uncovers the mathematical gem. The teacher–magician can do the same with one of the many number puzzles that produce a fixed answer. The students will be surprised to find that the "unknown" answer has been there under the paper all along.

6.4 Laboratory facilities in the classroom

In discussing the construction and utilization of classroom models, aids, and activities, emphasis has been placed on those that are readily available to the teacher, easy to make, and inexpensive. Likewise, in discussing laboratory facilities in general, only those readily applicable to the individual mathematics classroom will be covered. Ideally, each junior and senior high school should have facilities available for a resource center and for laboratory activities not only in mathematics but in each of the major subject-matter areas. Such facilities would offer space and materials for the construction of aids by both the teacher and the student, enough positions to handle a class working in small groups on different laboratory experiments at the same time, adequate resource material for independent student research, individual remedial-type materials, and a good collection of just-for-fun mathematical games and puzzles.

In a practical sense, however, most school systems simply do not have these facilities available and teachers have to adapt their classrooms as best they can to suit their own particular needs and interests. The teacher, textbook, and classroom laboratory facility thus complement each other to offer a wide range of abstract and physical mathematical experiences and free and guided discovery activities for the student.

Flexibility is a key word in the effective use of a classroom laboratory. First, the arrangement of the classroom itself should be flexible. Occasionally, desks should be arranged in small clusters for group experiments, or perhaps set aside in a corner of the room where selected students can regularly work on special projects. Second, the use of the materials available should be flexible. Certain units can be developed that make extensive use of laboratory activities. Others might involve an individualized learning situation in which the whole class or only certain individuals participate. At still other times the facility can serve to construct a model or check a reference immediately, on the spot, to clarify some point. Third, the extent to which a laboratory activity is organized and developed should be flexible. Even in the best homogeneous situation, different students with different interests and motivation will be able to pursue the same topic at decidedly different levels. Some students are adept at free discovery, while others will need guidance; laboratory activities should not be set into a rigid, fixed mold for every student.

Storage is often a problem in the classroom. Large, cumbersome, highly specialized aids should be avoided in favor of the more simple and versatile types. Materials should be kept in boxes in a closet or on a shelf or in a file cabinet in the classroom itself. There is no exact list of materials and aids that should be kept in each individual classroom, since there are so many factors that influence the choices. Subject, grade level, student ability, teacher interest, classroom size, and number of students are but some. Without attempting to be either too complete or too vague, the following suggestions are made.

Basic supplies

Of first importance are the basic materials needed for construction. These should include the following items:

> graph paper ruled in 1/10-, 1/4-, and 1/2-inch squares
> paper and tagboard in assorted colors
> acetate and pens for the overhead projector
> duplicating spirit masters
> pencils, markers, and chalk in assorted colors
> scissors, tape, glue, razor knife, staples, thumbtacks, toothpicks,
>   string, wire, rubber bands

Measuring instruments

The first two items listed should be available in classroom quantities. The others are useful and in some cases can be constructed by students. The last items can serve as the basis for interesting measurement experiments but do require modest funds for purchase.

foot rulers calibrated in both English and metric units

compasses and protractors

homemade clinometer or hypsometer to serve as a model for student construction

models of unit areas and volumes in both the English and metric systems

model of a linear vernier

100-foot tape

right-angle mirror

vernier caliper or micrometer

Computing aids

Computing aids can range from simple homemade devices to electronic pocket computers or teletype terminals. The following suggestions are for those with a very restricted budget. But even for those, some aspects of the modern electronic computer can be taught without the hardware itself. In particular, flow charting and the elementary computer language BASIC can be useful motivational materials for the classroom by themselves.

abacus with directions for student use

Napier's rods

slide rules in suitable quantities

nomographs of various types reproduced and ready for student use

Geometric models

A wide variety of commercial geometric models are available over a wide range of prices. Hence the choice of models depends to a large degree on their cost and the level of classes taught. The models listed should be found in each school, although not necessarily in each classroom.

geoboards in classroom quantities for junior high and general mathematics classes

1-inch cubes in reasonable numbers for individual experiments and for classroom demonstrations

set of models of the five regular polyhedrons and various prisms
and pyramids
cylinder, sphere, and cone of equal height and radius that can be
used to illustrate the volume relationship 3:2:1
cone that dissects to illustrate the conic sections

Mathematical games

Within the last few years a wide assortment of mathematical games has
come on the market. Some are primarily for fun; others relate to key
mathematical concepts; and still others are useful in reviewing basic arithmetic, geometric, and algebraic skills. It is virtually impossible to make
an exhaustive list of all those available. The games mentioned here merely
illustrate the various types commercially available and suitable for a laboratory in the classroom. They can be used in math club activities, played by
individuals during free time, discussed for the class by the better students,
and mastered for class-, grade-, or school-wide competition.

KRYPTO—This game is fast moving and challenging. Players gain practice in fundamental operations in arithmetic and in order of operations in algebra. Using any or all of the basic operations along with
five numbered cards dealt, each player tries to derive an expression
equal in value to a sixth card dealt.

(Available through Creative Publications, P.O. Box 10328, Palo Alto,
Calif. 94303)

POLYHEDRON-RUMMY—This game is designed to develop skill in
space perception through the use of cards containing polygons of
various shapes and sizes that can be arranged to form polyhedrons.
A winning hand is a set that forms one of the nine possible polyhedrons.

(Scott, Foresman and Company, 1900 East Lake Ave., Glenview,
Ill. 60025)

REAL NUMBERS—Directions for this game are simple and the game can
be used on a wide variety of levels. Natural numbers, integers, rational numbers, and real numbers are formed from selected numbers
and operations found by rolling dice. The objective is to form as
many numbers as possible within a given time limit. Operations include addition, subtraction, multiplication, division, powers, and roots.

EQUATIONS—Cubes with numbers and operations are used to build
equations while preventing opponents from building theirs.

WFF 'N PROOF—Involving cubes and cards, this set of games in logic can become quite complex. Useful for both the student and the teacher.

(The preceding three games are available from WFF 'N PROOF, 1111-HX Maple Ave., Turtle Creek, Pa. 15145)

HEXED—A set of 12 pentomino pieces to be arranged in a special way.

(Kohner Brothers, P.O. Box 158, East Paterson, N.J. 07407)

NUMBLE—Tile are placed on a board in special ways which involve computation with their numbered faces.

(Selchow and Righter, Bay Shore, N.Y. 11706)

QUBIC—A game of three-dimensional tic-tac-toe which requires good space perception. Interesting applications of this game to the teaching of coordinates are possible.

SOMA CUBES—Seven pieces that can be assembled in many ways to form a cube.

(The preceding two games are available from Parker Brothers Toy Company, Salem, Mass. 01970)

SPIROGRAPH—A fascinating set of gears of various sizes that can be used to construct many interesting geometric designs. Numerous applications to the mathematics classroom are possible.

(Kenner Products, 912 Sycamore St., Cincinnati, Ohio 45202)

Reference books

One excellent source of reference materials is the National Council of Teachers of Mathematics, 1906 Association Drive, Reston, Va. 22091. Some of its many pamphlets available at nominal cost and especially well suited for use by the mathematics student are given here.

Boxes, Squares, and Other Things: A Teacher's Guide for a Unit in Informal Geometry

Five Little Stories

Historical Topics in Algebra

How To Study Mathematics: A Handbook for High School Students

Introduction to an Algorithmic Language (BASIC)

An Introduction to Continued Fractions

Mathematics and My Career

Numbers and Numerals
Paper Folding for the Mathematics Class
Polyhedron Models for the Classroom
A Portrait of 2
Puzzles and Graphs
Secret Codes, Remainder Arithmetic, and Matrices
Topics for Mathematics Clubs

The N.C.T.M. also publishes pamphlets on the mathematics literature for the elementary, junior, and senior high school. The following books have been selected from these listings as representative of a basic collection of mathematics books that could find wide use by students in the classroom.

ABBOTT, EDWIN A. *Flatland.* New York: Dover Publications, Inc., 1950.

BAKST, AARON. *Mathematical Puzzles and Pastimes.* New York: Van Nostrand, Reinhold Company, 1954.

BELL, ERIC T. *Men of Mathematics.* New York: Simon and Schuster, 1937.

BERGAMINI, DAVID, and the Editors of *Life. Mathematics.* Chicago: Time-Life Science Library, 1963.

COURANT, RICHARD, and ROBBINS, HERBERT. *What Is Mathematics? An Elementary Approach to Ideas and Methods.* New York: Oxford University Press, Inc., 1941.

COURT, NATHAN A. *Mathematics in Fun and Earnest.* New York: The Dial Press, Inc., 1958.

CUNDY, H. MARTYN, and A. P. ROLLETT. *Mathematical Models.* New York: Oxford University Press, Inc., 1961.

EVES, HOWARD. *An Introduction to the History of Mathematics.* New York: Holt, Rinehart and Winston, Inc., 1969.

GAMOW, GEORGE. *One, Two, Three . . . Infinity.* New York: The Viking Press, Inc., 1961.

GARDNER, MARTIN, ed. *Scientific American Book of Mathematical Puzzles and Diversions.* New York: Simon and Schuster, 1959.

————. *Second Scientific American Book of Mathematical Puzzles and Diversions.* New York: Simon and Schuster, 1961.

HOGBEN, LANCELOT T. *Mathematics in the Making.* Garden City, N.Y.: Doubleday & Company, Inc., 1960.

HUFF, DARRELL, and GEIS, IRVING. *How To Lie with Statistics.* New York: W. W. Norton & Company, Inc., 1964.

JAMES, GLENN, and JAMES, ROBERT C. *Mathematics Dictionary.* New York: Van Nostrand Reinhold Company, 1968.

KASNER, EDWARD, and NEWMAN, JAMES R. *Mathematics and the Imagination* New York: Simon and Schuster, 1940.

KLINE, MORRIS. *Mathematics in Western Culture.* New York: Oxford University Press, Inc., 1953.

KRAITCHIK, MAURICE. *Mathematical Recreations.* New York: Dover Publications, Inc., 1953.

MEYER, JEROME. *Fun with Mathematics.* New York: Harcourt Brace Jovanovich, Inc., 1961.

NEWMAN, JAMES R. *The World of Mathematics.* New York: Simon and Schuster, 1956.

STEINHAUS, HUGO. *Mathematical Snapshots.* New York: Oxford University Press, Inc., 1969.

Activities

1. Read and report on the following chapters in the Thirty-fourth Yearbook of the National Council of Teachers of Mathematics, *Instructional Aids in Mathematics* (New York: N.C.T.M., 1973).

   Chapter 4, "Other Printed Matter," by Hilde Howden and John N. Fujii
   Chapter 7, "Projection Devices," by Donovan R. Lichtenberg
   Chapter 10, "Mathematics Projects, Exhibits and Fairs, Games, Puzzles, and Contests," by Viggo P. Hansen, Emil J. Berger, and William K. McNabb

2. Review a mathematical film or filmstrip and comment on its appropriateness for the grade level suggested, correctness of subject matter, quality of production, clarity of development, and number of concepts presented.

3. Prepare a set of 35-mm slides for use in the classroom on one of the following topics: Fibonacci sequence, applications of mathematics, or mathematics in art.

4. Prepare a lesson that makes use of a tape-cassette recording. Some suggestions are a historical narrative, directions for an activity that leads to a mathematical discovery, or an arithmetic speed test.

5. Write for a catalog from one of the suppliers of materials and models for the mathematics classroom. Then use it to identify and describe

several commercially available aids for geometry, indicating their potential use.

6. Find a source and price for a commercially made wooden dissectible cone that can be used to illustrate the conic sections.

7. Describe how the volume relationship 1:2:3 among a cone, sphere, and cylinder with the same radius and height can be developed through the use of commercially available models.

8. Prepare a flow-chart transparency that can be used to review skills in per cents. List various input and operations on movable strips of acetate.

9. Develop a transparency sequence that can be used when teaching the Pythagorean theorem.

10. Show how the equation of a straight line in slope-intercept form can be reviewed using the overhead projector and an overlay on a transparency grid.

11. Construct a transparency that can be used to review the formula for the area of a trapezoid, $A = \frac{1}{2}h(a + b)$. Show the steps to follow in a flow chart and use masked disks to select various values of h, a, and b.

12. Construct a 12 x 12 multiplication table on a sheet of acetate for the numbers 1 through 12. Use a mask with a square cutout to project only a 3 x 3 array from the table. Discover how the sum of the nine numbers projected can be found from the middle number in the array and then give an algebraic proof.

13. Prepare a bulletin board display entitled "Think Metric."

14. Collect at least 10 optical illusions suitable for a bulletin board display.

15. Describe how a bulletin board display can be constructed around a theme for a special holiday of the year.

16. Prepare a list of posters and charts on mathematics available from various organizations and industries.

17. Collect a set of drawings or pictures of famous mathematicians suitable for posting along with a brief description of the life and contributions of each.

18. Prepare a poster that illustrates a mathematical puzzle.

19. List five topics appropriate for a bulletin board display for each of the subjects: algebra, geometry, trigonometry, and probability.

20. Describe in detail a laboratory experiment involving a ruler, string, and some circular objects that students might follow in discovering an approximate value for π.

21. Identify several locus problems that can be easily illustrated with simple aids in the geometry classroom.

22. Cut out a rectangular piece of cardboard with length twice the width. Punch a hole in the center, at one corner, and in the middle of both a longer and a shorter side. Flip the rectangle along a straight line, side over side, and trace out the locus for each of the four points.

23. Describe how Activity 22 can be used to review coordinate geometry.

24. Discuss how a rectangular grid projected on a chalkboard can be used to teach congruency.

25. Prepare a set of cards with fractions, decimals, and per cents that can be ordered on the chalkboard by students.

26. Prepare a list of additional commercially available games suitable for the mathematics classroom.

27. Describe how the game of checkers might be used to introduce co-ordinate geometry in a mathematics class.

28. Make up a variety of games on mathematical skills that can be played with a simple deck of ten 3- by 5-inch file cards numbered 0 through 9.

29. Prepare a report on three of the publications of the National Council of Teachers of Mathematics that deal with aids and activities in the classroom.

30. Select one of the mathematics books from the list suggested on page 230. Report on its readability by a secondary school student and how it might best be used in the mathematics classroom.

References and selected readings

BEELER, NELSON F., and F. M. BRANLEY. *Experiments in Optical Illusions.* New York: Thomas Y. Crowell Company, 1951.

BOYS, CHARLES V. *Soap Bubbles.* New York: Dover Publications, Inc., 1959.

HARDGROVE, CLARENCE E., and HERBERT F. MILLER. *Mathematics Library—Elementary and Junior High School.* Washington, D.C.: National Council of Teachers of Mathematics, 1968.

JOHNSON, DONOVAN A., and C. E. OLANDER. *How To Use Your Bulletin Board.* Washington, D.C.: National Council of Teachers of Mathematics, 1966.

JOHNSON, DONOVAN A., and G. R. RISING. *Guidelines for Teaching Mathematics.* Belmont, Calif.: Wadsworth Publishing Company, Inc., 1972, Pt. 5, "Use of Teaching Aids."

KRULIK, STEPHEN, and IRWIN KAUFMAN. *How To Use the Overhead Projector in Mathematics Education.* Washington, D.C.: National Council of Teachers of Mathematics, 1966.

Mathematics Teacher. "Using Flow Charts with General Mathematics Classes" (Apr. 1971), p. 311.

————. "Visual Aids for Relating Direct and Inverse Circular Function" (May 1972), p. 431.

National Council of Teachers of Mathematics. *Computer Facilities for Mathematics Instruction.* Washington, D.C.: N.C.T.M., 1967.

————. *Instructional Aids in Mathematics* (Thirty-fourth Yearbook). Washington, D.C.: N.C.T.M., 1973, Chap. 4, "Other Printed Matter," by Hilde Howden and John Fujii; Chap. 5, "Other Media and Systems," by William Fitzgerald and Irving Vance; Chap. 7, "Projection Devices," by Donovan R. Lichtenberg; Chap. 10, "Mathematical Projects, Exhibits and Fairs, Games, Puzzles, and Contests," by Viggo Hansen, Samuel L. Greitzer, Emil J. Berger, and William K. McNabb.

NEWMAN, JAMES R., ed. *The World of Mathematics,* Vol. 4, Pt. 19. New York: Simon and Schuster, 1956 (paperback ed., 1962).

RAAB, JOSEPH A. *Audiovisual Materials in Mathematics.* Washington, D.C.: National Council of Teachers of Mathematics, 1971.

SCHAAF, WILLIAM L. *A Bibliography of Recreational Mathematics,* Vols. 1, 2. Washington, D.C.: National Council of Teachers of Mathematics, 1970.

————. *The High School Mathematics Library.* Washington, D.C.: National Council of Teachers of Mathematics, 1970.

INDEX